WINDOWS - DRAFT 3F
TITLE: WINDOW WALL SOUTH: WINDOWS 2nd FL.
SCALE: 3/4"=1'-0"
DWG.#: #W2Sc
DATE: 05/22/20
ADDRESS:
DESIGN: aardvarchitecture P.C.
37'-9½" (LOW ROOF)
28'-5" (4th FLOOR)
19'-6¼" (3rd FLOOR)
10'-7½" (2nd FLOOR)
0'-0" (1st FLOOR)
-2'-0" (BACK YARD)
2
2186 (86")
#2S-R
542
105
13
778
706
23
60
17
77
ABLE WINDOW (IN MM), EXTERIOR

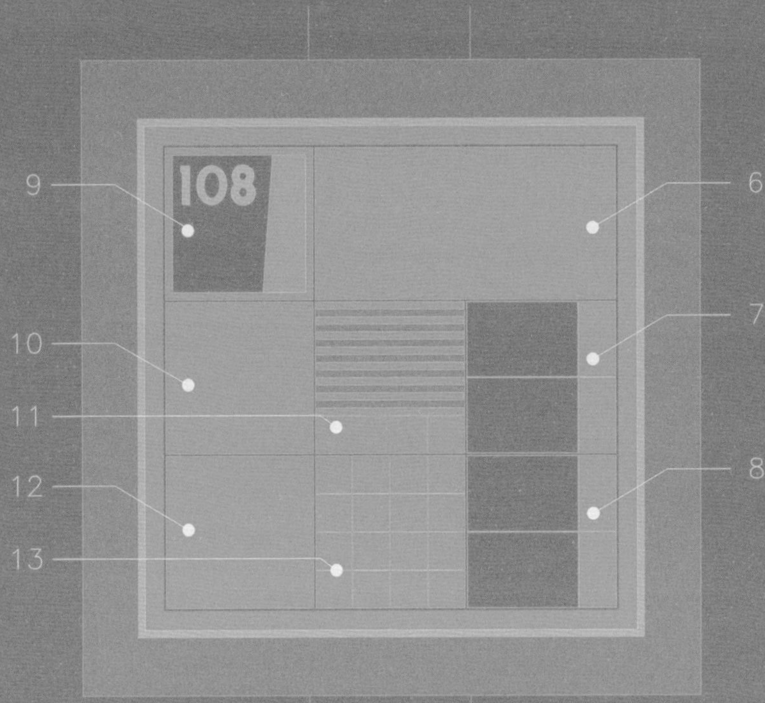

ELEVATION

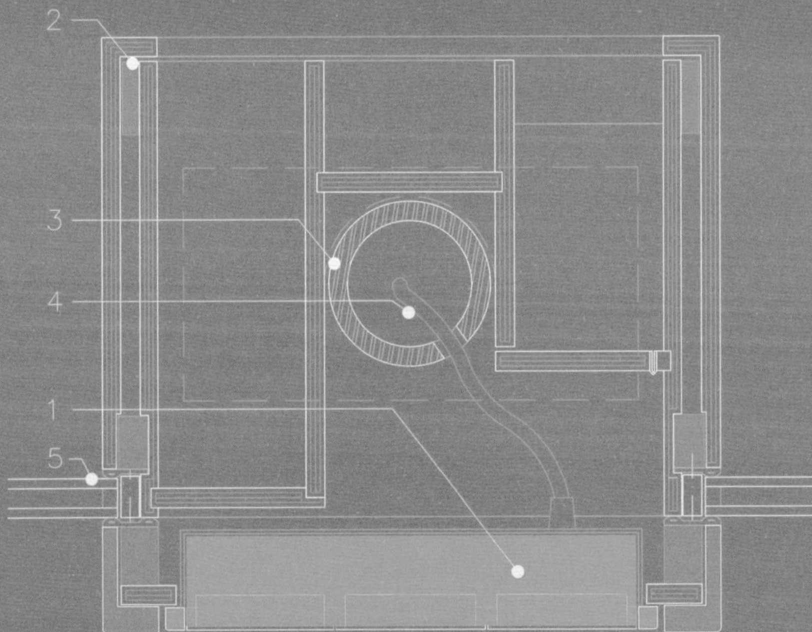

UNDERCOVER ARCHITECT

TOWNHOUSE DESIGN-BUILD IN HARLEM NY

CHRISTIAN VOLKMANN

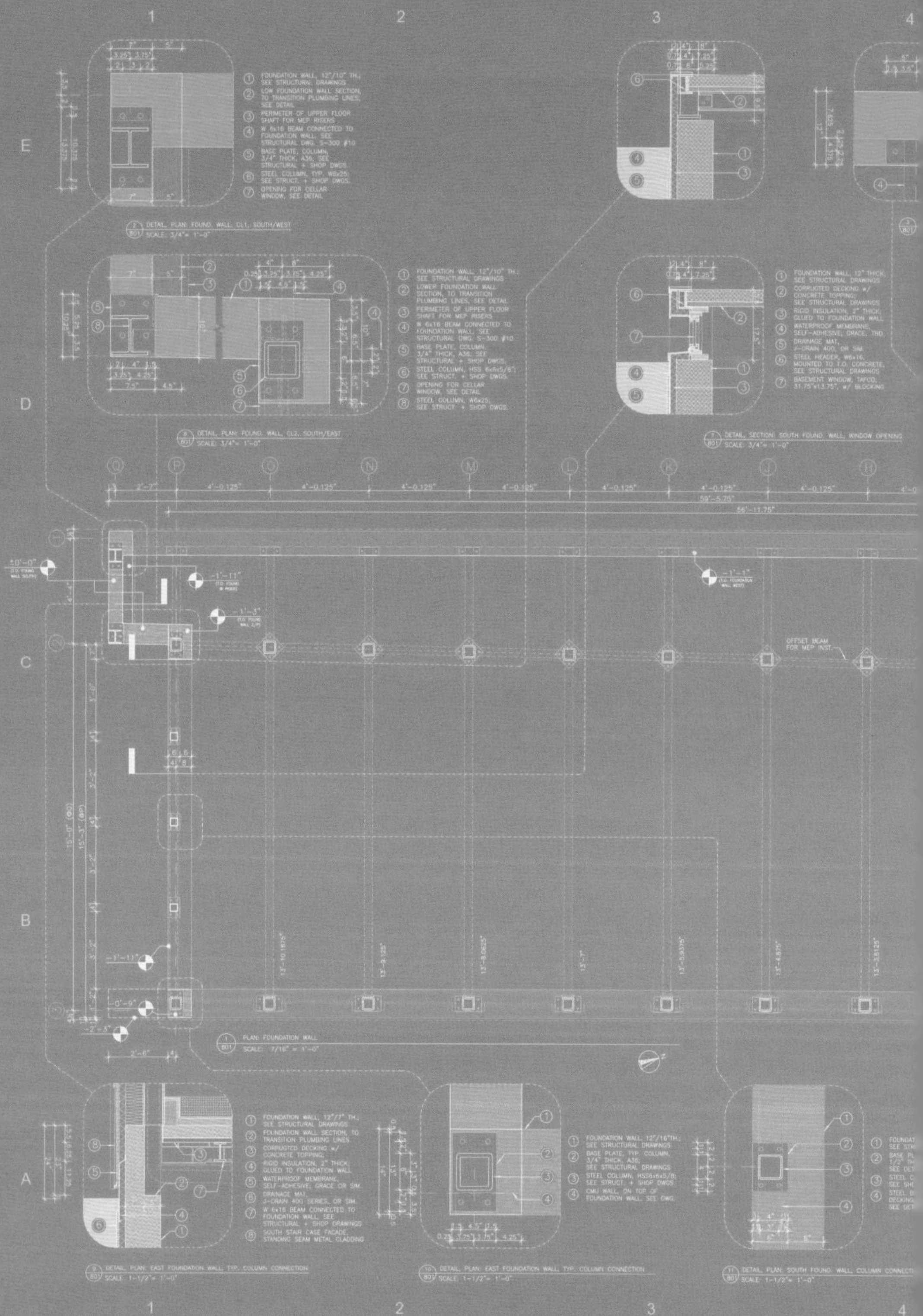

1 FOUNDATION WALL, 12"/10" TH.; SEE STRUCTURAL DRAWINGS
2 LOW FOUNDATION WALL SECTION, TO TRANSITION PLUMBING LINES, SEE DETAIL
3 PERIMETER OF UPPER FLOOR SHAFT FOR MEP RISERS
4 W 6x16 BEAM CONNECTED TO FOUNDATION WALL, SEE STRUCTURAL DWG. S-300 #10
5 BASE PLATE, COLUMN, 3/4" THICK, A36, SEE STRUCTURAL + SHOP DWGS.
6 STEEL COLUMN, TYP. W6x25; SEE STRUCT. + SHOP DWGS.
7 OPENING FOR CELLAR WINDOW, SEE DETAIL
2/801 DETAIL, PLAN: FOUND. WALL, CL1, SOUTH/WEST
SCALE: 3/4"= 1'-0"
1 FOUNDATION WALL, 12"/10" TH.; SEE STRUCTURAL DRAWINGS
2 LOWER FOUNDATION WALL SECTION, TO TRANSITION PLUMBING LINES, SEE DETAIL
3 PERIMETER OF UPPER FLOOR SHAFT FOR MEP RISERS
4 W 6x16 BEAM CONNECTED TO FOUNDATION WALL, SEE STRUCTURAL DWG. S-300 #10
5 BASE PLATE, COLUMN, 3/4" THICK, A36, SEE STRUCTURAL + SHOP DWGS.
6 STEEL COLUMN, HSS 6x6x5/8"; SEE STRUCT. + SHOP DWGS.
7 OPENING FOR CELLAR WINDOW, SEE DETAIL
8 STEEL COLUMN, W6x25; SEE STRUCT. + SHOP DWGS.
8/801 DETAIL, PLAN: FOUND. WALL, CL2, SOUTH/EAST
SCALE: 3/4"= 1'-0"
1 FOUNDATION WALL, 12" THICK; SEE STRUCTURAL DRAWINGS
2 CORRUGTED DECKING w/ CONCRETE TOPPING; SEE STRUCTURAL DRAWINGS
3 RIGID INSULATION, 2" THICK, GLUED TO FOUNDATION WALL
4 WATERPROOF MEMBRANE, SELF-ADHESIVE, GRACE, THD
5 DRAINAGE MAT, J-DRAIN 400, OR SIM.
6 STEEL HEADER, W6x16, MOUNTED TO T.O. CONCRETE, SEE STRUCTURAL DRAWINGS
7 BASEMENT WINDOW, TAFCO, 31.75"x13.75", w/ BLOCKING
7/801 DETAIL, SECTION: SOUTH FOUND. WALL, WINDOW OPENING
SCALE: 3/4"= 1'-0"
OFFSET BEAM FOR MEP INST.
59'-5.75"
56'-11.75"
1/801 PLAN: FOUNDATION WALL
SCALE: 7/16" = 1'-0"
1 FOUNDATION WALL, 12"/7" TH., SEE STRUCTURAL DRAWINGS
2 FOUNDATION WALL SECTION, TO TRANSITION PLUMBING LINES
3 CORRUGTED DECKING w/ CONCRETE TOPPING;
4 RIGID INSULATION, 2" THICK, GLUED TO FOUNDATION WALL
5 WATERPROOF MEMBRANE, SELF-ADHESIVE, GRACE OR SIM.
6 DRAINAGE MAT, J-DRAIN 400 SERIES, OR SIM.
7 W 6x16 BEAM CONNECTED TO FOUNDATION WALL, SEE STRUCTURAL + SHOP DRAWINGS
8 SOUTH STAIR CASE FACADE, STANDING SEAM METAL CLADDING
9/801 DETAIL, PLAN: EAST FOUNDATION WALL, TYP. COLUMN CONNECTION
SCALE: 1-1/2"= 1'-0"
1 FOUNDATION WALL, 12"/16"TH.; SEE STRUCTURAL DRAWINGS
2 BASE PLATE, TYP. COLUMN, 3/4" THICK, A36; SEE STRUCTURAL DRAWINGS
3 STEEL COLUMN, HSS6x6x5/8; SEE STRUCT. + SHOP DWGS.
4 CMU WALL, ON TOP OF FOUNDATION WALL, SEE DWG.
10/801 DETAIL, PLAN: EAST FOUNDATION WALL, TYP. COLUMN CONNECTION
SCALE: 1-1/2"= 1'-0"
11/801 DETAIL, PLAN: SOUTH FOUND. WALL, COLUMN CONNECTI
SCALE: 1-1/2"= 1'-0"

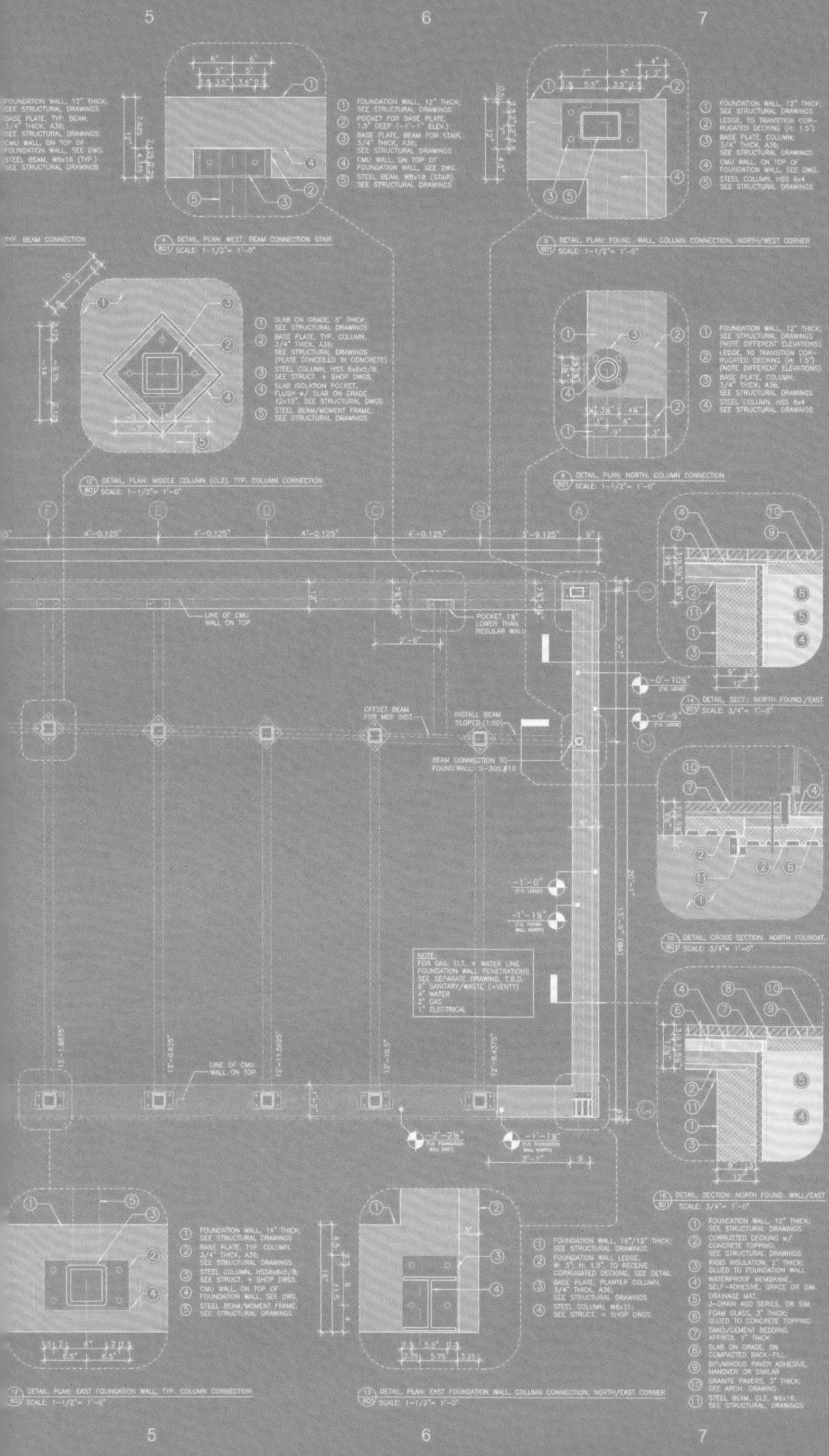

ALTERATION TYPE 1:
NEW CONSTRUCTION/FAMILY RESIDENCE

DATE: SUBJECT:

LOT NO.: 001
DRAWN BY: CV
CHECKED: CV
DATE: 06/01/2018

CONSTRUCTION SET

TITLE:

PLAN:
FOUNDATION
WALL SET-UP

SCALE: 7/16" = 1'-0"

PLAN-NO.:

A-801.00

PAGE: 0x OF xx

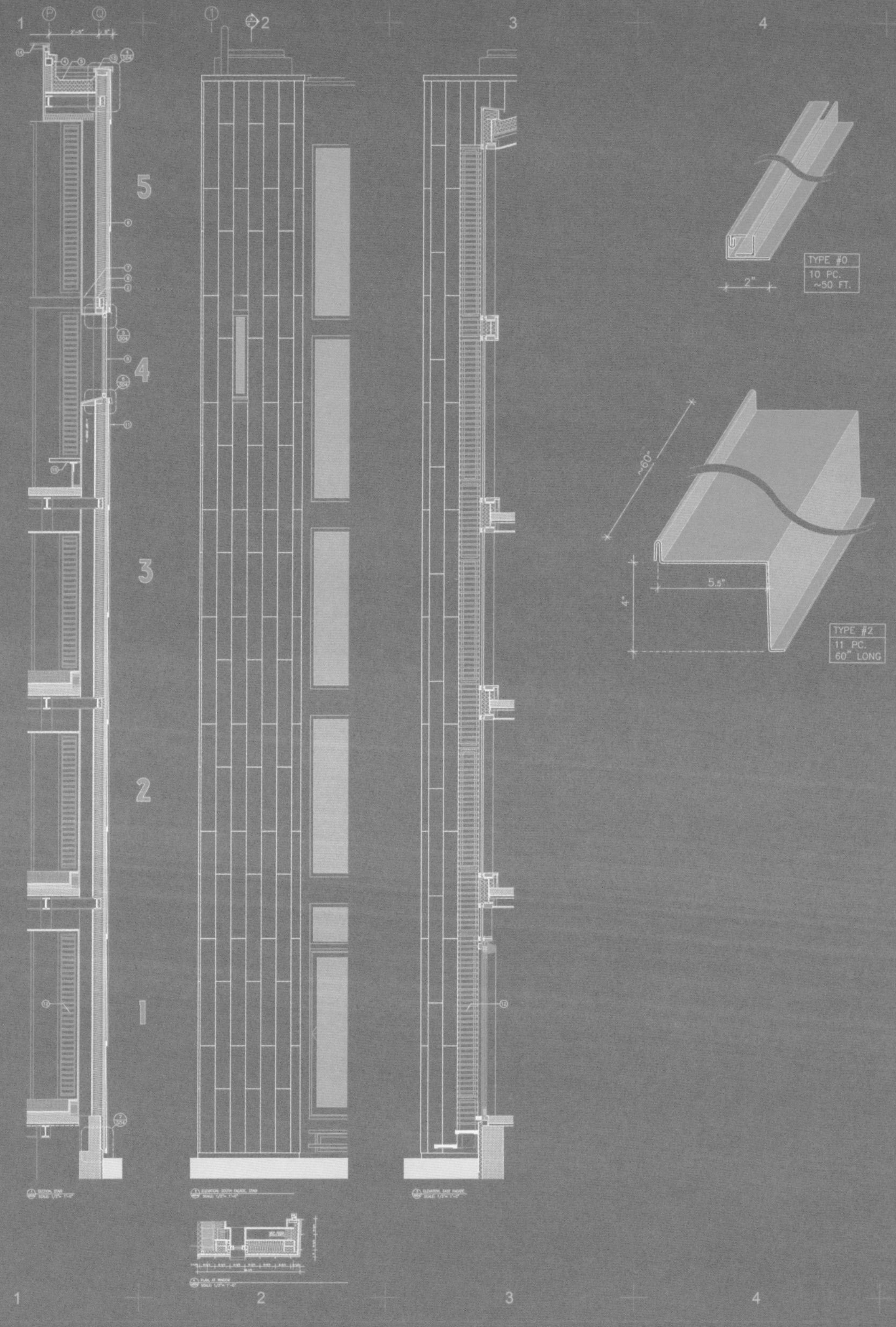

1
2
3
4
5
TYPE #0
10 PC.
~50 FT.
2"
~60"
5.5"
4"
TYPE #2
11 PC.
60" LONG

UNDERCOVER ARCHITECT

TOWNHOUSE DESIGN-BUILD IN HARLEM NY

CHRISTIAN VOLKMANN

OSCAR RIERA OJEDA
PUBLISHERS

TABLE OF CONTENTS

INTRODUCTION

When I was 17, I decided to become a carpenter. It was only after I told my parents about this and recognized their disappointment *("why did we send the boy to high school?")*[1], that I started to consider the alternatives.

Architecture had always interested me quite naively, yet I felt the strong pull from the *magic-of-making*. From today's perspective, I believe that few architects really know what they get themselves into when they decide to pursue this profession. Maybe only the ones whose parents were already associated with the profession, happened to know the true job description.

One of my first recollections in the pursuit of architecture came in part from reading the novel: "Billiards at Half-Past Nine," by the Literature Nobel Prize laureate Heinrich Böll. In Germany, a well-known story about processing the guilt after the second world war, it featured a family dynasty of architects spanning three generations. My fascination for the profession, as expressed in my native German as *Baumeister* (aka.: *Masterbuilder*[2]), was focused on the evolution of generations of architects in the story. The first-generation architect always engaged in meaningful conversations with the trades, and seemed to know as much as the trades people about their craft. Yet he was always one step ahead, and maintained the vision and idea of the artifact. Knowing and communicating the bigger picture all the while understanding the diverse *techniques* of building was an enticing proposition. During my design studies, this connection was again reinforced by one of my favorite teachers who compared the architectural profession with that of an orchestra conductor. *"You have to have knowledge of the possible techniques an instrument can play in order to come up with an idea of sound."*[3]

Architecture is an *applied art*, and can be defined as "*the only activity that involves imaginative conception of ideas, leading to artifacts that are realized as actual constructions ...*"[4]. However, it has become less clear in what sense the architect contributes in translating their *imaginative conception* into *actual construction*. It seems that the mastering of both *designing* and *building* has been lost. It also seems that the set-up of common operation has become so bureaucratized that the immediate connection between the two realms is difficult to accomplish, at least not with today's common know-how. Returning to the earlier analogy of the conductor, we have to understand what instruments are to be played and in which

"An Architect is a mason who has learned Latin."

Adolf Loos (Vienna, 1930 at "Grundsätzliches")

way, in order to protect and realize the special skill we were trained to offer. This understanding must be constantly calibrated to employ necessary processing tools in an increasingly complicated world. But how?

In New York City, real estate is one of the three largest economic sectors. There is always potential to capture a greater share of value in this *large-scale* market. Yet, why is so little of the built environment architecturally exciting? To be clear: not all construction aspires to be *applied* or the result of *imaginative conception*. It is evident that the majority of new construction in New York is neither seriously imaginative nor meaningfully concerned with beauty.

What's particularly troubling is that architects, who once set out to change the world through visionary ideas, have become complicit in feeding a system that treats buildings as commodified products. In a capitalist market, optimized production creates value; the more standardized the output, the more profitable the enterprise. The dominant objective becomes *return-on-investment* maximization.

There seem to be widely accepted mechanisms at work that conflict with the passion that first drew us to architecture: curiosity, inspiration, and the pursuit of tangible, meaningful experiences. These are the qualities that form the cultural value of our built environment. Architecture, as the enduring witness to the spirit of its time, appears less concerned today with artistic ambition. Stranger still, while the majority of architectural production follows this trend, a small percentage of showcase projects, often designed by famous European architects, operate under a different set of conditions, which often require extravagant budgets. Though we share a collective fascination with beauty, in practice we accept that architectural outcomes often underperform, stifled by a range of mundane and constraining factors. We accept this, it seems, because we believe we must.

Yet I've long suspected that we, as architects, could do much better, if only we let go of certain preconceived ideas about how the system must function. Instead, we should precisely observe and adjust our strategies to the way the current conditions play in our favor. Returning to the conductor's example: it seems we've lost both our inner sense of sound and our conceptual clarity. We've distanced ourselves from *technique*, and we neglected to immerse ourselves in the fundamental processes of architectural production. If architecture is truly an *applied art*, then perhaps it is also a form of *applied thinking*. Can *applied thinking* be taught? Can it be observed, or meaningfully communicated? Much of our learning comes through personal experiences, and through the reflective process of understanding how specific lessons fit into the broader world. This book shares my experience of building a townhouse in Harlem, a project that required me to redefine my role as an architect. But it is not merely about one project. If this experience can be extrapolated to suggest a more meaningful, integrated way of practicing architecture, assuming that similar trends and conditions continue, then perhaps I can contribute something of value to the profession: a personal (and at times unconventional), method of *applied thinking*.

Hence the title of the book: "*Undercover* (= my own experience) *Architect* (= all of us)."

To distinguish between different modes of writing, the book uses two different page colors throughout the book: the white background presents general observations about our multifaceted profession, while the gray background shares anecdotal reflections and specific stories from the process of designing and building the house in Harlem. These narratives are grounded in first-hand experience and relate directly to applied architectural practice. In this book different structures overlap. The images in the book are in chronological order, however, the chapters can't be. Each chapter focuses on certain knowledge, based on what I consider the most important players of architectural construction. In a design-build environment, you might be playing different parts as one person. The resulting dissonance creates narrative tension and offers the reader a puzzle to solve.

My Story: A Townhouse in Harlem:

A few years ago, frustrated by the lack of opportunity in my professional life as an architect, I sensed that something had to change. Fueled by that frustration and, admittedly, by a certain stubbornness, I decided to put a new premise to the test. An opportunity arose, and I seized it:

I chose to run my own construction firm and develop a new townhouse from the ground up, right in the heart of Harlem. To prepare, and to make the project as economically viable as possible, I trained in multiple disciplines. I first qualified as *Construction Superintendent*, a certified position required by the NYC Department of Buildings. I coordinated all building trades, commissioned all subcontractors directly, and maintained full control over scheduling and logistics. I also served as my own subcontractor, working hands-on across many trades: steel structure, concrete work, framing, drywall, tile, stone, cabinetry and even daily site cleanup. I didn't mind the physical labor, it was a financial necessity at first, but over time, I realized that the architect should be closer to construction responsibilities. That proximity

001

enhances understanding and enables better control of design execution. Beyond construction management, I also took on financing, budgeting, and code compliance. As the *Architect of Record*, I handled the entire design process, from initial concept through construction documents. I managed much of the administrative work as well, far beyond the typical scope of an architect. What began as a financial necessity, gradually became a preferred mode of operation. I came to understand that true control over design requires serious day-to-day hands-on participation.

Throughout the process, I wore many different hats: designer, builder, manager, and laborer. This experience gave me a new perspective on integrating design and construction, and on developing a more cohesive, strategic practice.

I share this story to encourage fellow architects who may be seeking alternative methods of project execution. To gain control, we must accept that our profession is too multifaceted to be confined to a single job description. *Designing*, which we often consider the heart of our work, is not enough on its own to bring an architectural vision to life.

THE DEVELOPER

002

003

004

What is a developer? Can you study such a profession at College? According to my recent internet search, developers "*oversee a project from start to finish: selecting a site, securing the finances, overseeing design and construction, and managing or selling a property. The developer, sometimes with the help of a builder, is responsible for forming a team of professionals to implement his or her vision within a certain budget and schedule, all the while following applicable codes, ordinances, and policies*."[5] This is kind of surprising for me, because it sounds like the developer is competing with the architect's profession and skills set, maybe with the exception of managing or selling a property (though I know architects who do this too). When starting my education as an architect, I considered our profession to be the one with the *vision*, and thus the one *overseeing the project from start to finish*. So, it makes me wonder what the qualifications of the developer are that haven't also been addressed in our professional training. For example, architects have unique team building and communications skills that should be considered a major asset to owners and builders.

Inherently, architects are in control of the schedule and budget as well. We also must be knowledgeable about codes, ordinances, and policies: In fact, we (as licensed architects) are the ones who sign and seal the necessary documents with the building department. On paper, we carry the responsibility, but financially we do not. Maybe that's the crux of the matter: The loss of responsibility happens because developing and building involves spending substantial amounts of money. Do developers and owners see us as merely the *artist*, and thus not the right person to be trusted with significant amounts of money? There are lots of projects where architects exceed their budgets exorbitantly in the process of the construction phase, often due to the passion they have for their product, and because they seem to have limited understanding of the *means and methods* of construction and how they affect cost. Architects often do not look for the most obvious solution, but instead they process the world in a more visual and spiritual way, looking for something curious and inspiring. Therefore, often the ideas they come up with, are not easy to create, resulting in work that is more difficult and more expensive to construct.

With developers, this overspending doesn't happen as easily. The developer is more focused. The developer knows how to estimate the costs of a project more prudently from experience, or they may even utilize a *cost estimating consultant* or a

005

006

007

value engineer. One of their main objectives is about generating profit. The architect (in the common setting), on the other hand, is less beholden to the future the income derived from the project. It has been my suspicion for the longest time that we, as architects, could do a better job by wearing both the *creative hat* and the more pragmatic *thinking cap,* and thus conquering more ground in the field. We have not been very eager to accept risk as part of the profession and, consequentially, have been taken less seriously as team players in the construction field where collaboration and problem-solving is key to producing good results. Architects have lost the power of being recognized as the logistic coordinators. We are too focused on our idea generation and drafting tasks and have lost the connection to what is tangible and that which defines the architectural experience. If we focus on learning more about the coexisting activities of constructing architecture in order to produce a better comprehensive product, we could again become competitive.

It seems, the developers are conquering the vacancy that is left by the lack of readiness from the architectural community. It is this emerging discipline that focuses on built architecture as a tangible result, not as just being processed on paper. The advantage of the developer is that they hold the *business knowledge* to run this kind of operation, and that they are used to carrying most of the financial risks.

However, what the developer is often lacking (except the very good ones), is the imagination for a more versatile and more sophisticated product. Often, developers are using a predictable *cookie-cutter* approach for their projects. The result is that our built environment often has little surprise or elevated experience for the users. Projects led by architects are often more considerate and inspiring in terms of context, space, and detail. To combine those assets with logistical management skills would lead to a better overall product and to a more design-literate and appreciative society. I realize that the way I am explaining this is rather *black & white*. There could be more differentiation. But maybe it is also a question about naming the profession: if we play our cards right, what once was called a *Masterbuilder* (see introduction) and later was renamed the *Architect*, might perhaps become the *Design-Build Developer* when the skills of the architect are fully utilized and optimized along with the traits of a developer. That said, I care much less about the title, and more about the qualitative output and the comprehensive skills associated with this challenge.

A long time ago, my partner and I started thinking about purchasing a lot in Harlem, New York. The stock market was very hot and the competition and demand for acquiring real estate in Harlem was low. So in 2001 we became the proud owners of a 20' x 100' lot, and thus, without knowing it, developers-in-training. The dedication and the disposition were there, but the more I understood what needed to come together in order to build a house on this lot, the more I realized that this was easier said than done. In the beginning, my mind was purely that of the architect, thinking about how to organize an interesting architectural layout. I studied all different kinds of influences: how daylight hits the site, how the cornice heights of the adjacent buildings would align, and other things. I visited the property frequently and worked, especially during weekends, on removing rubble and garbage that people had dumped on the site. I enjoyed it. Increasingly, I became enchanted with the beauty of this site and its context. It was a stunning block with many assets: churches, community gardens, and many street trees. The property taxes were initially low. Had I known about the current *property tax levels* (rates which are now fifty times higher than when we purchased the lot), I may have reconsidered taking on the same challenge today. However, at that times, expenses were low and the future looked bright. I was determined to put a new building up on the site.

008

009

010

The first time I recognized that my way of thinking about the project had significantly changed was when the developer of the adjacent lot on the East side (where a decrepit carriage house from a century ago, damaged by fire, was awaiting demolition) asked for a meeting. We spoke at length, and discussed the logistics of his upcoming construction project (he hoped and planned to use our lot as his staging area). I realized the questions he asked were focused on operation and logistics, and there was no detectable desire to produce beautiful architecture. I initially didn't understand most of the questions and so my responses were likely inadequate. But afterwards, I thought more about the challenges ahead, and started to develop a different and more pragmatic perspective on the project.

After meeting with the adjacent property developer, not much happened for a long time. The other adjacent building on the west side was so fragile (with many concerning cracks in its demising wall) that it would have been irresponsible to attempt putting a new foundation next to it. The building changed ownership a couple of times, and in early May 2012, it famously collapsed on a

011

012

013

014

015

016

017

018

019

020

021

022

Friday afternoon at 4pm (with no casualties). After the unexpected collapse, I took one step closer to becoming a developer. Now the land was worth much more, and it was now feasible to construct a building on the lot. Despite the opportunity and the favorable conditions, it still took a long time to get the project underway. On the one hand, I had my constant doubts (and I was often reminded of them, when I spoke to architect friends of mine) about being knowledgeable enough to come up with sufficient construction strategies. On the other hand, I was nervous about the long bureaucratic process of getting the project approved and ready for a building permit. This was even more difficult, because the neighborhood was now zoned into a *landmarked "Historic District,"* meaning that approval from the Landmarks Preservation Commission (LPC) would be required before the Building Department would even consider looking at the drawings. Since I wanted to build in a contemporary *style* (which is a misnomer, when one is actually living in a contemporary world), I knew there would be objections. To further add complexity, the local Community Board had to be involved and issue a *"letter of acceptance"* prior to the Landmarks Preservation Commission's review. That meant countless hearings and discussions, until acceptance was finally granted. Many property owners of *brownstones* on our street were concerned about contemporary architecture. They preferred consistency to what they were used to seeing (even though that was built under conditions more than a hundred years ago). Their suspicion was somewhat understandable, because they had experienced promises from past developers that were given to get to the next step, only to have these promises broken with subsequent legal methods. The built results were not convincing. It helped however to have been part of the neighborhood for a long time, and I had friends and neighbors who cared and supported me in such hearings and discussions.

In 2018, I was finally ready to start. However, by that time, the owner of the adjacent Western lot (with the former building, now collapsed) had changed yet again. I met with the new developer for lunch, and we got along well. He actually offered to buy my lot, because it was clear that building one larger building over two lots would be more profitable than one narrow building on his own lot. I politely declined, and explained to him as good as I could that I was really longing to *build*. He had sympathy for me, and in time, became in a special way, a mentor of mine. On many future occasions, he advised me what to do and gave me crucial pointers for driving the project forward. I now was learning as a developer.

023

024

025

At that first meeting, we promised each other to *help-one-another.* We also discussed how to use the same contractors for the initial work, again, because it was less expensive for both of us to do so. A couple of months later, after I got my building permit issued by the Department of Buildings, I kindly asked him to sign an *access agreement* to his lot, because I was eager to start. An *access agreement* is necessary to obtain from the adjacent neighbors for excavation, because the many types of excavation often concern and involve the adjacent lots in an urban setting. Excavation can be extraordinarily costly and complicated. They typically require a 45° *angle of repose*, which means that you would have to encroach any adjacent lot in immediate contact with the future cellar walls, in order to build your foundations. My neighbor didn't like the idea of granting me access before he was ready to start his construction site. His building permit was still pending. He also understood that starting the adjacent construction sites together would actually lead to more efficient progress, amid the obvious cost savings.

Instead of explaining to me that this is not the right move, making his own excavation more difficult to follow, he articulated this idea: He would issue the access agreement for a payment of $50,000 if I really wanted to start early. Upset about the exorbitant number that I didn't really expect, I asked him what had happened to our mutual agreement to *help-one-another?* He explained that, in his eyes, we were doing exactly that with this amount of money: If I wouldn't have access to his lot and still wanted to start early, I would have to pay for *shoring* for the entire 100 feet of adjacent property line, which would approximately cost me about $1,000 per foot or $100,000 for the 100 feet depth of the lot. That $100,000 dollars would instead be saved by sharing the profit with one another: $50,000 of savings for me and $50,000 for him. I am not sure how serious he was about this proposal. He was, of course, more experienced than me in the world of development and I believe that he just wanted to test my reaction.

When I was first employed as an architect, my boss explained to me what he called the *"magic triangle of construction management."* [6] The three sides of the triangle represented *time, cost,* and *quality*. He finished his initial explanation with the observation that in *real life* you can only satisfy *two-of-the-three* criteria. This is an interesting concept, especially for architects who want to be involved in construction. It is clear that wanting to stay true to a conceptual idea is making *quality* the top target. How an architect defines *quality* differs from how a developer understands it. For the

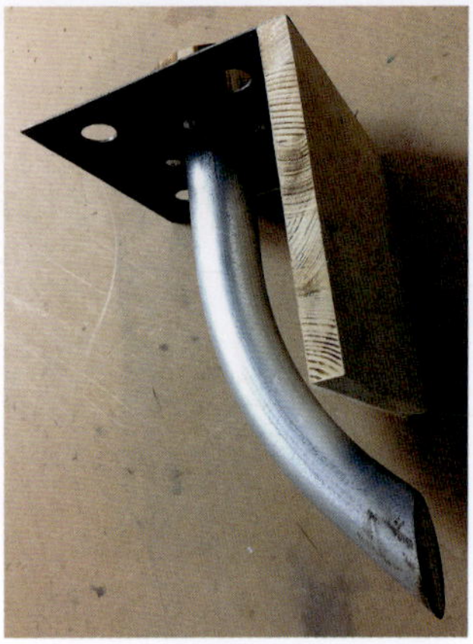

026

027

028

029

030

031

032

033

034

035

architect, *quality* relates to a particular *vision* and how a particular design concept is manifested in its *architectural* detailing. For the developer, *quality* relates much more to repeatable craftsmanship and clean execution of construction. Therefore, the developers' choices are broader, and they can be dictated by the related *costs*. For the architect, there is less flexibility, even though this problem must be resolved as well, since budgets are most often limited. Therefore, for the architectural mind, often *time* is the most flexible criterion, involving the amount *time* to solve the problem with proper detailing and how it can be executed. The developer would seldom sacrifice *time* for *quality*. Here, the flexibility is between *quality* and *costs*. Solutions are found that are *acceptable*, even though they compromise coherence and intent of the architecture. Both mindsets are understandable. However, the architect is economically disadvantaged. *Time is money.* Therefore, skills to expediently resolve issues while maintaining architectural quality are instrumental in making progress and in being competitive.

What concerned me most (as a developer) was the budget. I started by creating meticulous spreadsheets to predict various expense scenarios and formulate a project budget. Cost estimation services exist, but given my intimate knowledge of the project and their propensity for estimating too high, I did the calculations myself. The final number hovered at around $1.5 to $2 million (which is an intimidating number). Compared to a dollar per square foot ratio (which developers are constantly monitoring, in order to assure profitability), it sounded very reasonable: *costs* of $350 to $400 per square foot were what developers at that time were targeting. In order to hit those numbers, the architectural detailing had to adjust to what was possible but still looked beautiful. I thus had to contribute to the quality of detail in a much more physical way, which was my personal involvement in the construction. I have been *good-with-my-hands* throughout my life. Not only could I save money by self-performing work, I knew that it would also result in more architecture and better control. This is a different strategy any developer would suggest: Their strategy is often to calibrate input in labor with the quality and cost of the outcome. They therefore like to repeat the same detail as often as they can, especially if it contributes to the enhanced value of the building. This is not necessarily bad, because it leads to consistency in the proposed architecture over time. It also leads to lower prices, because the subcontractor, after a while, will know much better what to expect. But it is a little bit like eating the same food everyday ... it is boring and it leads to a lack of *vitamins*. In any case, the decision-making is different: The developer favors less conceptual content, but more what is *in style*.

My budget concerned me so much that I hired a *condo lawyer,* who would be able to render a condominium plan. This is a common way to navigate from one project to the next: You first subdivide the property formally if you know you do not have enough money for the project. Then, you promise to sell one (or all) of the apartment units. The *return-on-investment* will be there, because whatever it costs to create an apartment is likely less than what it then can bring for sale in the New York City real-estate market. This is again architecture functioning in the realm of a capitalistic market. The idea is not the architectural asset itself – it might not be detected by the potential buyer. The architectural system functions in favor of market forces: What sells is what is *en-vogue*. Seldom will people figure out independently what they appreciate besides the predominating perceptions.

036

Often, the first question to be asked is about how many bedrooms a unit has. To me, in terms of processing architectural culture, this is wrong: people, not involved in architecture, should be able to appreciate an architectural idea beyond the economic mainstream. Counting the bedrooms originates from real-estate advertisement and considerations of *resale-value*, a word I learned in the first month of living in New York. I took a risk in trying something special with the layout of the units. Instead of pushing for a design to maximize the amount of bedrooms, I decided to provide one continuous *loft* space per apartment, thinking that the way we live has changed (i.e. less than 20% of households in Manhattan are traditional families[7]) but also because I experienced the spatial beauty of the lot beforehand, with an exciting relationship between the rear gardens and a church building on the front street side. In German, this configuration is called *Durchwohnen* which translates to *through-living* and is characterized by access to daylight from two sides and also improved cross ventilation. Many friends, also architects, warned me about the unconventional floor organization. Later on, after the building was completed, I discovered that many more people than I expected fell in love with the *identity* of the space. I admit that the units are not very suitable for families with kids. But in New York, many different markets exist. It was inspiring that many people who visited the apartments left with enthusiasm, because they had seen something they liked and they did not expect. That enthusiasm is what architecture must connect to. The problem I (and my friends) anticipated didn't exist. There is demand for good architecture. Many people don't just like what is promoted. They

037

038

039

040

041

were able to discover something new and identified with it. There is hope for creative thinking and thus for architects to stimulate peoples' architectural senses. In the end, my condo plan didn't work out. The lawyer failed to follow-up on the tasks crucial for making progress, and on the contrary, was asking me to provide the work that I expected him to do. He also struggled to see the extra value in what he considered just a transfer of property.

I was lucky that during that time I was also in contact with a small bank. They inspected the property a couple of times. They finally offered me a loan that was in hindsight better than giving up some of the *equity* of the building. I went ahead with the loan. I got through the closing procedures and I learned what it means to finance a piece of real estate in New York. In time, I also learned how to deal with payment schedules for the subcontractors, and how to coordinate with subcontractors in order to push for progress. I learned how to submit tedious applications (i.e. for electricity, gas, water) and I became more familiar with the intricacies of dealing with permits to be obtained from the Department of Building.

In the end, all units became rentals. While this means that I will carry the burden of financing for a much longer time, I still maintain full control of the building as an asset. I will also be able to do the typical developer task of *managing the property* on a long-term basis, again with lots of new things to learn.

042

043

044

045

046

047

048

049

050

051

053

052

THE ARCHITECT

054

055

056

It is crucial *where* you start your architectural career. The formative academic environment you encounter will affect your understanding of how architecture fits into the world and how the world influences architecture. This particular mindset will follow you for life. I started my architectural education in Germany and Switzerland. At my universities, there wasn't a distinct divide between the profession of the architect and that of the engineer. In fact, in Germany even the official academic title for a student graduating in architecture was a *Diplom-Ingenieur (Dipl. Ing.):* an *ingenieur* is an engineer. It was expected by the architect to be (a bit of) an engineer, and able to comprehend principles of engineering. To understand the profession as a technical endeavor and to embrace the *means and methods* of construction in order to fit into an interdisciplinary construct. What was interesting in my particular environment was that the same was also the case for the engineering degree: there was emphasis given on being aware that architectural objectives are different than those of engineering. In fact, some exams were taken jointly by both architects and engineers. The academic exchange was intended to strengthen both professions and encourage interdisciplinary collaboration. This makes a lot of sense in many contexts, from academic research to the implementation of construction: when starting a construction site none of the things initially visible are yet related to architecture. At first glance, pretty much all of it is related to engineering with masses of raw building materials and dirt. The engineers and the architects, even though they focus on different things, must interact and understand the different perspectives in order to successfully complete a project.

When I came to the United States and starting working at an office (amid my obvious hands-on tendencies perceived by my colleagues), I was advised that as an architect one should *"never get caught with a hammer in ones hand."* [8] A distinct differentiation was made between the active builder on a job site *(blue-collar)* and the draftsman in the office *(white-collar)* whose job it is to document something on paper without any sweat or tears. However, how can you draw something competently if you haven't been exposed to the way it's actually put together? There is a distinct pool of *know-how* that is supposed to inform both fields, but it seems to be neglected by separation and differentiation. The architect therefore seems very isolated in their position because of the

057

058

059

strange niche *we* created as a profession. *We?* Who really created it?

The *American Institute of Architects (AIA)* confirms the intended isolation in one of their popular contract templates, the *Contract for Construction (A201).* Within this document it is stated, *"The Architect shall not have control over, charge of, or responsibility for the construction means and methods ... The Architect shall visit the site at intervals appropriate ... to become generally familiar with the progress and quality of the portion of the Work completed, ..."* [9/10] It practically denies the architect meaningful influence of their own design and creation at the jobsite. It also cuts short any viable source of learning and interaction with builders on site. The reason is clear: a lot can go wrong on a construction site. Over time, the architect (for whatever reasons) decided to reject more responsibilities related to construction, avoiding any risks of liability in favor of rather being a bystander. By waiving this responsibility, the architect loses influence during the most critical time of the genesis of the building.

To isolate architects from experiencing and understanding the *means and methods* of construction deprives them from a learning process that is essential in an *applied art* profession. The techniques of how something is to be constructed requires an adjustment to our way of thinking. Our mindset must relate to that of the experts in the field. Therefore, communication with craftsmen is essential to grow a repertoire of *techniques*. To withdraw from the opportunities of learning from each other, means that we are essentially withdrawing from the field of *making*. Closer contact with the building trades is needed in order to successfully lead the charge in the creation of sophisticated architecture. (The equivalent is an orchestra conductor who doesn't know how to play an instrument; or, the maitre-du-chef, who refuses to cook.) During my time in Germany, I have experienced other models that position professional architectural work in a much more advantageous way: There, for smaller buildings, the architect is generally responsible for choosing the subcontractors and tendering the related contracts. This leads to ongoing business relationships, and often to better execution and higher quality of work. It is clear that the collaboration over years and decades leads to more mutual understanding and to more refined and coordinated detailing. Both entities support one another in the process of

060

061

062

making and understand the effect that a particular detail must achieve in the context of *applied art*. There are famous architects, such as the Italian architect Carlo Scarpa, who is said to have worked with the same craftsmen for decades [11]. I would argue that the result of some of the most beautiful architecture credited to Carlo Scarpa is actually a serious collaboration between unknown craftsmen (who didn't seek to get recognized) in the way Scarpa, the architect, was longing for it. These craftsmen knew how to utilize their skills into a direction that Scarpa would find commensurate to his particular *visions* of architecture.

Architects are said to have big egos because historically they have often wanted to claim all the credit related to a project for themselves. However, it is clear that the expression of the building in terms of its craft is based on good teamwork, between craftsmen and designers. A good architect is appreciative of the input he or she receives from people who have worked in their particular building trades for decades. It might be that the way of thinking of both entities is different. Scarpa, for instance, described it this way: *"We are not interested in mediocrity, we already know beauty, we search for the sublime."* [12] Establishing the overlap between *technique* and *sublime* expression is crucial for becoming a good architect. Only with good communication skills and deep knowledge can this exchange be established.

Scarpa's statement is important for this chapter. In the past couple of pages, I have described the necessity for the architect to get closer to engineers and craftsmen. However, the architect's perspective is still different. Besides being able to understand technical requirements, the architect's job is to *search for the sublime*. Beyond the rational understanding, the architect is orchestrating the built environment in a way that it becomes meaningful and, one could even say, transcends a bit into a world of magic. Otherwise, you shouldn't call it an *applied art*. Rather, it is related to a materialized experience, which is and should be beautiful. The artifact therefore benefits from the other involved stakeholders understanding the experience trying to be achieved by the architect (again, the same could be said for the equivalent of the maitre-du-chef and the orchestra's conductor). What is even more interesting is that Scarpa speaks of *"we"* and not just of himself. It appears that he is including the craftsmen in the *search for the sublime*. He wants

063

064

065

066

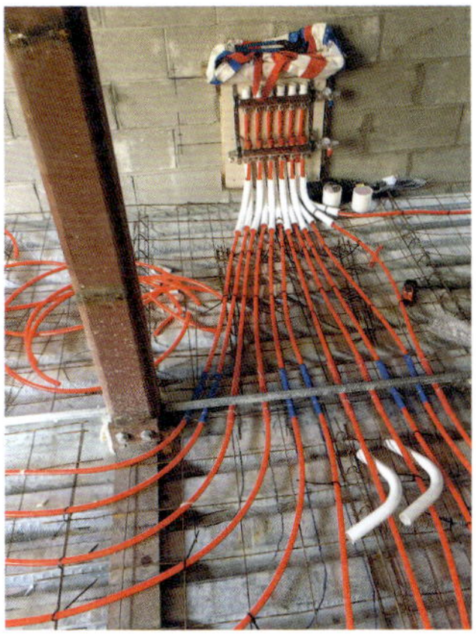

067

068

069

070

071

his craftsmen to understand the intended beauty with the eyes of the architect. Those seem to be the ideal conditions: Everyone can understand each other's *language*. However, a recent conversation with a general contractor I invited to one of my reviews at my college explained (in front of my students) that, as a general contractor, you do not want an architect on a job site, because it is very time-consuming and the workers are less likely to get anything done. For that particular reason, from his perspective, it made sense: the general contractor must be most concerned with meeting deadlines and maintaining the schedule. The longer construction tasks take, the less it'll be in the interest of the contractor. Maybe this is just a cliché: I am curious what conversations took place at his job sites between architects and craftsmen. It is clear that the *language* did not align. If it was time-consuming, maybe it was because none of the parties involved knew how to relate to one another. If so, who is at fault? To me, it is clearly the architect. We should take and have the responsibility of communicating efficiently if we want to be in control of the *process*. But how do we get there? Good communication skills only develop with exposure to the tangible tasks at hand. If we have no or limited knowledge about these tasks, we are left on the sidelines of communication.How can we learn from a place, the construction environment, that doesn't want us to be there.

Maybe the architect's big ego is a counter-reaction to not being taken seriously in the context of construction. However, if there is no risk-taking in the profession, how much should the construction team appreciate the input of the architect? If the source of the architect's creative thinking is muted on the construction site, then that original *sublime* intent is seldom related to the final result. It is certainly not visible to the craftspeople who have worked on it. The *firewall* the architect puts up after the project is finished to establish that the entire idea was his or hers is completely flawed and counterproductive. It is clear that in architecture, the tangible idea went through many hands and is the result of collaboration and teamwork. The main skill of becoming a good architect is about bringing the positive energy of *sublime making* together, and being the skillful and attentive conductor in the process. The architect's pride must thus be shared, and of course, the architect can only be confident if the process of comprehending is instilled in the output of others.

072

073

074

Before construction on my townhouse started I was in constant touch with many of the builders I had worked with before. Some of them I knew for quite some time. For example, the steel structure (which is instrumental for setting up the entire geometry of the building), was done by the father of a former student of mine who I trusted and had worked with about ten years prior. After the structural engineer issued the preliminary drawings, we were in frequent contact during the planning phase as well as before fabrication began. Several details were changed based on these conversations, shaped by years of experience, and then again evaluated by the structural engineer. I ended up creating the shop drawings myself, as I had a good understanding of the structure and the details based on my discussions and conversations with the fabricator. I also visited the steel shop regularly and became familiar with typical techniques and details by observing and conversing with the steel workers.

075

I grew close to the steel workers' mindset and to the responsibilities of getting the assembly as right as possible due to the affect this trade has on subsequent construction. I participated for weeks during the installation of the steel structure. I learned things I would not have learned in any architecture office. Later during the project, I was also able to self-fabricate certain pieces that were crucial for the building because of my experience with the steel workers. With the stairs, for example, I started by creating plywood stair stringer and tread/riser bracket templates. I used the wood template to cut the final steel stringers and fabricated them according to the templates. Throughout the process, I spent a lot of time at the steel fabricator's shop. Being invited into their community was both fun and rewarding. The experience has also changed how I draw and reminded me that the people who come with the perspective of *how-to-make-things* can fundamentally change my way of thinking.

076

Another close friend of mine (who I have known for more than 20 years) has been a craftsman in many different contexts. He was instrumental in how I think about myself within in the world of architecture, confronted with many challenges of logistics. Architects are very demanding, often without knowing what steps it takes to produce their *visions*. Formerly an installer for mechanical systems, he decided to run his own *fine-cabinetry* operation which has now been in business for three decades. He is very detail-oriented and he does not take no for an answer. Understanding that architects think differently from craftspeople, he always asks questions and is not shy to communicate his approach to solving a particular problem from his perspective as a cabinetmaker (but more often as a general *tinkerer*).

077

078

079

080

081

082

083

Some of this attitude I adopted for my own habits at the construction site. In his observation, many architects cannot think through a problem beyond their own horizon. The solutions they propose are deficient, and therefore they rely on the craftspeople in the shops and in the field to fix their issues. Even worse than not being able to help solve issues arising during the construction of their projects, many architects are incapable of anticipating the potential problems before they are discovered. This is not surprising, since we seem to avoid shops and construction sites, where such skills are obtained.

Initially, I found myself taking the side of the architect, insisting that the drawing must be right and the fabricator must follow and adjust to the proposed conditions. My friend, on the other hand, was trying to show that much of the burden is on the side of the architect who lacks technical anticipation. He was often so passionate about the topic based on the consequence that financial penalties during construction are most often with the fabricator, who is under pressure and must find ways of making things work in a practical way. The architects, however, are more financially insulated from this type of damage. In most cases, they do not bear any financial consequences, and that affects their temperament.

Many architects, from his perspective, are not able to contribute to a meaningful conversation because they cannot think and problem-solve like a fabricator. This especially frustrated him because most of the problem-solving to bring something beautiful into existence (which was his intention as well) falls in his court along with the financial risks. Since we were friends, we soon both realized that putting the blame on one another is not the way to deal with such logistical shortfall. The real issue is that communication between the parties is often harmed by the way the system is fundamentally set up. Usually, the architect does not talk to subcontractors and fabricators because they are separated from one another by a General Contractor. However, exposure to fabrication and *making* should be one of the main sources of learning for the architects who are creating the drawings in the first place. I realized that this learning experience (which I instinctively had a tendency for) is instrumental to produce drawings that are proficient to drive the process. I therefore started to embrace many of the architectural detailing problems as if I were a cabinetmaker, or steel worker, or mason, etc. An architect's contribution in terms of thinking and idea-making cannot just exist on paper. It must be brought to life with the same passion for tangible beauty.

Throughout construction, my friend advised me on how he would try to solve certain issues. This affected how I approached the challenges

084

085

086

with my own thinking and it influenced me greatly. His extensive experience for dealing with technical issues shifted my way of problem-solving to something that was much better at balancing the *sublime* world of the architect with the more pragmatic world of the craftsmen.

Sometimes the conversations we had were eye-opening, and other times they just confirmed what I already thought (which was reassuring). Other times I found his input to be frustrating, because it seemed as if there was no mutual solution to be had. This then forced us to both think harder until we got to a point of agreement where our thoughts overlapped. Throughout the project, I realized that I was adjusting how I found myself processing architectural problems. It is a good lesson for the architecture profession in general: don't just carry around your own insecurities without having a chance to reflect them with the insecurities of others. Communication is key for learning (for all parties), especially for disciplines that depend on comprehensive input. It leads to a change of *building culture*. The project required many other craftspeople, whom I didn't know beforehand. My adjusted way of thinking helped to nurture collaboration and trust. To accomplish this, the craftspeople had to be treated in a way such that they knew that I was always available to help solve problems with them together. They had to know that I would not abandon them with issues that had the potential to cause logistical or financial hurt. These new relationships were more challenging, because every party in this business is vulnerable, and therefore more suspicious. Again, spending *contact hours* either on the phone or ideally in person, was crucial. It is important to firmly communicate that problems will be solved together, without any risk of spending significant additional money or labor on their part. The issue with communicating efficiently is that the architect is often perceived as not equipped to have such discussions. Furthermore, the system is set up such that they usually go through the General Contractor middleman. All of this makes it more difficult to solve any problems. And, in consequence, it might have repercussions on the schedule. Despite of what the *AIA* contract articulates, we can indeed come and go to the construction site whenever we think it is productive for the work to be done. It leads to a more efficient *process*: to be on time and on budget.

What concerned me early in the project was the realization that a building cannot rely on beauty alone; it must also function effectively. The primary challenge in this *parti* was designing the mechanical and plumbing systems. A specific complication came from the long, continuous stair that ran the full depth of the building. This spatial condition made installing the building systems particularly complex. The design approach, enticing in its spatial qualities, is actually not very

087

088

089

090

091

092

common anymore in the developer's world, because it requires more attention to the detail: specifically how to coordinate space, structure and all the runs of piping and conduits comprehensively.

Typically, a single main riser is located next to the vertical stair core and runs straight down to the cellar. However, that approach was not possible here. After many discussions, the mechanical engineer developed a clever solution. Understanding his proposal required advanced spatial and system thinking, as it involved tracing the main *arteries* of the system in a non-traditional way. We had worked on it together with many sectional sketches, before the final solution was found. However, during several jobsite visits with plumbers interested in the project, it became clear that the design's theoretical intent on paper was not being understood when discussing the scope of work. Many of the plumbers simply could not grasp the unconventional concept. To address the lack of understanding on site, I took steps to make the installation process more intuitive. Before the concrete was poured, I created a system to embed all the necessary sleeves directly into the slab. I built them using foam-filled PVC pipes, so that the intended layout would be self-explanatory during construction. This method also helped reduce costs. Because the sleeves were pre-installed, no core drilling was required to create openings through the floor slabs. This was a significant saving, as drilled penetrations would have involved many hours of labor and expensive equipment, such as diamond-tipped drill bits.

The approach also helped eliminate risk. I was especially concerned about core drilling because the heating system (which I had also installed myself) used a hydraulic radiant heat system. Any damage to the pipes embedded in the concrete floor during drilling would have been a serious problem. To prevent this, I established a strict policy: no core drilling was allowed without my explicit permission. On site, I kept a binder with extensive photographs documenting the location of all radiant heat pipes in the floor before the covering concrete was poured, so I could assess the risk of damage if any changes were proposed. (That said, I was once overly confident, and a pipe was damaged. It took me an entire day to carefully remove the surrounding concrete and repair it.) Many subcontractors laughed at the idea of using a radiant heat system. Even though it is far more efficient in performance and it avoids unsightly radiators, typically below windows, they claimed it would inevitably be damaged during construction and therefore it wasn't worth the trouble. But with a great deal of supervision, I was able to protect the entire system, and it remained fully intact through the process. Despite this success, I still couldn't find a plumber willing to take on the job for a reasonable price.

I decided to add something else to the jobsite after recognizing what was causing the hesitation: I physically drew the available space above the ceiling of the stairwell for the plumbing and mechanical installation directly onto the adjacent demising wall using a permanent marker. It took me half a day during a weekend. To articulate this in space, not on paper, was critical to making the system work. The following day, I invited the plumber I hoped would take on the job. As soon as he saw the lines, he understood the intent immediately. We spoke for another twenty minutes. During our conversation, he also identified an error in the original concept and proposed a solution. We then discussed in detail how the design could be adjusted to accommodate it. He sent me a reasonable quote the next day. It was clear I would hire him, as he had grasped the scope of work better than I expected and showed confidence in both the design and the hours involved. This moment marked a major breakthrough in the project.

093

Trying to communicate the same concept with the drawings was unsuccessful. Without the permanent marker lines, effective communication would not have been possible. This experience showed me that you have to adapt your communication strategies intelligently to match the *language* to the way of thinking of the plumber. Doing so opens doors and helps reduce costs. Why does it save money? Any subcontractor who feels unsure about the exact scope of work will likely include significant overhead in their proposal, simply out of concern that the installation may be more complex or take much longer than expected.

094

Another example, which occurred much later in the construction process, involved applying stucco to parts of the staircase. As mentioned, this was the most important space in the building, as it connects all the spatial experiences. In this case, the necessary communication centered around mutual trust: trust in each other's skills and reaching a consensus on how things should be done, balancing the craftsperson's *technique* with the architect's spatial and material intentions. Typically, the architect is not involved in such decisions, since *means and methods* are managed by the general contractor. But this situation was different. First, I was the general contractor overseeing the site. Second, the subcontractor, someone I had known for twenty years, had a rather unconventional way of training his apprentices. One moment stood out and left a lasting impression on me: Before I even spoke, the subcontractor addressed his crew with the following: *"The architect here has a vision. We have to do our best to fulfill that vision. Before we discuss how to do this, we will listen to him."* That was a powerful statement because it created real alignment between the builders and me as the architect. I explained that both

095

096

097

098

099

100

101

102

103

104

105

106

107

the risers of all steps and the adjacent exterior demising wall were to be finished with a continuous layer of stucco. The goal was to create a unified brown/grey surface that would then visually counterbalance a second material, white in color, to generate a specific rhythm throughout the space. Here, in fact, a rendering was instrumental to explain the intended effect. I emphasized how important it was for the stucco to appear seamless, without any interruptions in its surface. This was easier said than done. We developed a plan for how the work needed to be carried out on a wall that was 60 feet long, varied in height (up to all five stories at the entrance), extending diagonally through the building. Teamwork was essential to ensure that the curing process would not compromise the quality of the work. Some workers were assigned to the mixing station, while others worked in pairs to apply the stucco – first the scratch coat, then the finish coat. We figured out how to conduct the process together, step by step. In the end, the workers proudly said: *"We were responsible for translating the vision."*

It was a heart-warming moment because it revealed a genuine commitment to working together on something intended to become beautiful. The workers were not just focused on completing their physical tasks; they also respected and contributed to the intentional beauty of the project. In exchange, I learned to understand the challenges of the hands-on *processes*. Once the work was finished, they were as proud of the result as I was. Although I initially found the word *vision* somewhat questionable, since it sounded as if I had been under some kind of influence, I later realized the term might have been more appropriate than I thought. As architects, we are trained and specialized in *vision*, understood as visual recognition. We think differently because we rely on visual reasoning to imagine how architecture should appear and how it might produce beauty. This relates to Scarpa's search of the *sublime*. My hope is that our visual skills give us our sense of beauty, which can lead to something truly special. The challenge is that none of these skills are visible before any work begins, making it difficult to convince others in advance that the effort is worthwhile. Still, how can we be convincing if we are absent from construction sites and unfamiliar with the *processes* required to bring that vision to life? We must contribute on site if we want to become literate in the practical steps that lead to tangible architecture. The previous story was encouraging to me because it showed that we, as architects, can offer real value on site – especially if we focus on how to communicate effectively with the builders, who are ultimately the ones tasked with interpreting the *vision* in material form.

As an architect, you have to engage in efficient communication. You must be a master communicator, beyond your own horizon. The most

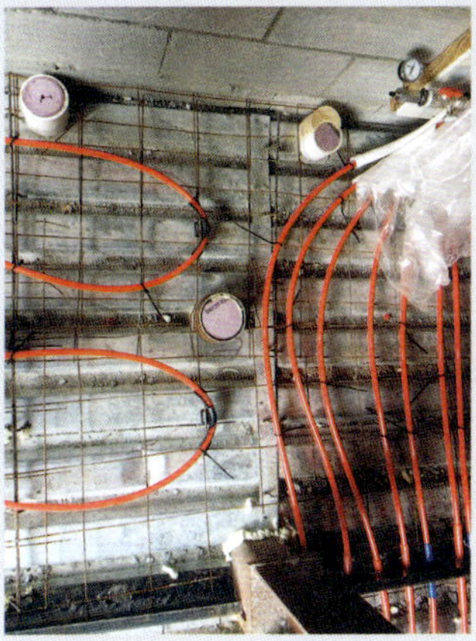

108

109

110

important part of communication does not actually happen through the drawings. In many cases, the drawings serve as legal documents, rather than as tools for meaningful dialog. Rarely do they lead to the exchange that clarifies the goals of creating beautiful architecture. There is a vast amount of untapped knowledge held by craftspeople who have worked for decades. It is troubling that there is no clear platform to access and learn from that experience. Perhaps it is not even possible in a formalized way. Often, this learning of applied material knowledge is shared informally, through conversations that arise from curiosity and a willingness to spend time together and to ask questions.

Being taken seriously by the workers on a construction site is essential to an architect. Without that respect, there is no meaningful exchange and no real control. Earning that recognition also strengthens my confidence in the *vision* that others expect me to create. It suggests there is still a path back from the detrimental tendency of architects to remain uninvolved and uninterested in the construction process. Architects should not believe in the strict separation of roles among stakeholders in construction, because this mindset harms our profession the most. The reputation of the *masterbuilder* can only be restored if we use our communication skills to gain applied material knowledge. Only then can we begin to understand the *secret language* spoken in that environment. The same is true in other contexts, such as financing and development. We must be able to speak many different *languages*. Only then can our drawings begin to carry real power. Planning is tied to understanding, and understanding means learning how to adapt to the *language* of others.

111

112

113

114

115

116

117

118

119

THE BUILDER

120

121

122

Everyone is drawn to good craftsmanship. In order to be a good architect, you must genuinely appreciate it. Ludwig Mies van der Rohe once said, "in the detail the eye of God is present."[13] But the detail does not come solely from the architect's drawing. It also emerges from the builder's experience, and from the architect's ability to observe and learn from good craft.

It becomes a kind of *the-chicken-or-the-egg* question: What comes first? The craftsperson's detail or the architect's drawing that anticipates it? In the end, one must admit that both forces were likely once unified in a single individual. How can we, as architects, hope to anticipate the craftsperson's detail in our drawings if we don't engage deeply with both the physical work and the finished appearance?

The best way to learn is to be proactive in conversations with talented and skilled builders. Even better is to engage in informal, *ad-hoc* apprenticeships. By this, I mean taking the initiative to pick up a tool and offer real help on the construction site. *Helping* requires sensitivity to the perfection and pride that craftspeople take in their work. They want to produce something of quality. If you are unaware of that, you risk getting in their way and disrupting the *process*. But if you observe closely, you can learn and eventually, should be able to assist meaningfully. If the craftsperson and architect were once one figure, they might become one again. It is up to us, as architects, to engage in learning the handicraft that is in support of architecture as an *applied art*.

The skill sets among builders in New York are incredibly diverse. At some point, you must trust them, but you also have to be careful about who you trust. Who is a good craftsperson? Who is a good teacher? Who takes pride in their handiwork? There are outstanding individuals, who care deeply about the quality of their work and have accumulated rich, practical knowledge over many years. There are also others who are more focused on business efficiency, prioritizing speed and output over precision. The spectrum is wide. It is not always easy to tell who is who, but it usually becomes clear through early conversations during setup and planning. These discussions about *how-to-do-the-job* often reveal a person's agenda, level of thoughtfulness, anticipation, and pride in the outcome. What is most surprising is that, in some construction trades, there is no formal education at all. For example, if you want to become a cabinetmaker in Germany, you must complete an apprenticeship under a master cabinetmaker – a title that is officially certified.

123

124

125

That apprenticeship usually lasts three years, though some are even longer. It covers the essential skills needed to become proficient in the trade. This includes not only how to fabricate items specifically to that discipline, but also how to run a business. As a result, the average quality of work is far more predictable. Conversations about the work tend to find common ground, since everyone shares an understanding of the complexities through a standardized education.

In New York, by contrast, if you decide to become a cabinetmaker, you can start working almost anywhere, and you could remain at the same level for decades. The assumption seems to be that the market will determine your success: if your work is good, you will thrive; if not, your business will fail. But it is not that simple. Where do you even begin? The environment and the exposure to knowledge vary widely. Your growth often depends on how much attention your superior is willing to give you. The best builders I know are those who are committed to passing their knowledge to the next generation. The more patience and care they show in their teaching, the higher the quality of their craftsmanship often is. For architects, communication opportunities with builders are often limited. This has to do with how construction sites are typically organized. As an architect, you are expected to communicate mainly with the general contractor. This severely limits your exposure to the actual builders and to opportunities for learning from the people who execute the work. You must find ways to work around this system and approach craftspeople directly to learn about *materials and methods*. It is not just about what you want to communicate to others. It is about what you can observe and draw out from those who work with tools in their hands. You must be prepared to learn and observe.

My curiosity about craftsmanship and communication is what drives my learning. I am always longing to understand the builder's perspective. Doing so not only builds trust and teamwork, it also allows you to immerse yourself in a different mindset – one rooted in *making*. It is not about black lines on paper, which demonstrate your competencies, but about mastering real materials, which are often more stubborn and unpredictable than expected. Every material has its own strategies for being handled well. As architects, we often lack this knowledge because we were not trained to get our hands dirty.

If you are unsure how to get access to this knowledge, consider signing up for a summer internship with a local contractor. At first it may seem like the

126

127

128

people on a construction site speak a different *language*, but it is not difficult to pick up on the meaning. Once you begin to understand it, you can apply what you learn directly to your future drawings. It will make you more fluent in that *language* and it will help you understand the *means and methods* contractors use to keep a project moving efficiently. If you are fortunate to find a good mentor, it will open your eyes to how architecture and construction are truly connected. One of the key lessons you will learn is that drawings are less significant on the job site than strong communication skills and a good sense of empathy and anticipation.

However, you must be considerate in how you approach this kind of learning. In the world of construction, time is of the essence. This is understandable: A builder takes on a specific scope of work for a fixed fee, which must be estimated accurately. If the estimate is too high, they will not get the job. If it is too low, they may make no profit or even incur a loss. Every project is, in that sense, a financial gamble. Because of this, builders place high value on efficient time management and reliable quality output. Misunderstandings that lead to costly mistakes must be avoided at all costs. This is why *"pre-task meetings"* are important. These typically take place each morning and are led by the *General Contractor* or the *Construction Superintendent*. In addition to safety procedures, the meeting must clearly define the upcoming work scope and exactly how it should be carried out. Miscommunication at this stage is a frequent source of error. Even if things seem calm on the surface, there is often tension behind the scenes as everyone pushes to ensure a smooth start to the workday. The architect is usually absent from these meetings, even though they are the ones who hold the clearest understanding of how they want the result to look. Here the disconnect becomes crystal clear between how work should be done and what it should look like. The builder is carrying the risk of any mishaps, which will only be known after the fact, when the architect is inspecting.

This is precisely why architects should be present at such instrumental meetings. Many misunderstandings could be avoided if architects led these meetings and explained how the *means and methods* connect to the intended experiential outcome of the design. The nervousness that comes from the involved financial risk could be eased by clear and confident communication directly from the designer. If this happened earlier in the process, before estimates were even submitted, it would not only create a more collaborative and respectful jobsite culture – it also would lead to meaningful cost savings.

129

130

131

132

133

134

135

136

137

138

139

140

General contractors often say they do not want architects on site because they are concerned it will waste time and involve too much small talk. At my job site I was both the executing *General Contractor* and the inspecting *Architect*. It was clear that the stakes were high to communicate with the right attitude. Architects have to learn to speak a *language* that is concise and that meets the mindsets of the builders.

The best conversation I have ever had with a builder began with my assumption that you must choose between focusing on *quality* or on *speed*. My tiler, who has always been extremely precise, corrected me. He said that from his perspective you actually need to do both. He explained that he had to let workers go because they either failed to meet his quality standards or, even if they did, they took too long to get to complete their tasks. According to him, delivering excellent work within an acceptable time frame requires a particular kind of talent. Skill and time management must go hand in hand. He put it simply: "Not everyone can play the violin." That made sense to me. You need to bring a certain level of innate talent just to get started. And, as musicians, improvement in building often comes through refining *technique*. Developing the right *technique* enables you to be efficient and precise.

This same builder also spoke about his early love for chemistry. He said he often thinks about it while working with cementitious materials. The best admixtures, he explained, are discovered through trial and error. To make good progress you need to get the mix's consistency just right. But that consistency changes as the chemical reaction occurs. Many concrete workers describe their findings in similar, empirical terms. They refer to the mix in everyday language – comparing it to the consistency of toothpaste, pancake batter, or cookie dough. I have heard many of these descriptions, which sound more like something from a cookbook than from a technical manual. Of course, the instructions are only helpful, if you have some experience with the material yourself. If you have never tried to make the *recipe*, you won't really understand what it means.

You have to understand the principles of chemical reactions in order to understand the *sweet-spot* for action, whether you are applying stucco or pouring concrete. This kind of understanding is also empirical. I have heard many amusing stories from craftspeople about their early learning experiences. None of this is visible in drawings, and none of this material knowledge can truly be taught without direct hands-on exposure. Sometimes, the architect must learn like a chef, through experience, to understand what is possible with materials in the context of real, tangible architecture.

It is also important to recognize the differences in how people think. Architects tend to focus on the final result and the broader context, thinking in terms of the *big-picture.* Craft-oriented builders, by contrast, are much more focused on details and *process*. They are often completely absorbed in the specific task at hand. If the architect wishes to communicate effectively on site, they must learn to understand and adapt to this way of thinking.

Another experience that taught me a great deal about material behavior, particularly concrete, was both early in the project and extremely nerve-racking. My contractor, who had completed the foundation earlier, was relatively flexible about pouring the slabs over the installed corrugated steel decking. Normally, concrete is poured during warmer weather, when it cures faster and allows the surface to be finished the same day. However, steel work took longer than expected, and we were delayed. Before pouring, the radiant floor heating system also had to be installed. I taught myself the necessary skills and completed most of that installation on my own. The process took time, and by the time we were ready, the third-floor slab was scheduled to be poured in January. Temperatures were in the low thirties, even on the best day in the forecast. This meant the concrete mix had to include calcium chloride, an accelerator that prevents the water in the mix from freezing during the curing process. Earlier in the fall, when we poured a slab in October, we were able to walk on it within several hours. But January was a different story. The concrete simply refused to set. The workers waited patiently, long after the regular workday had ended. They attempted to smooth the surface, but that step can only be done once the concrete is firm enough to support the person doing the finishing. By 8:00 pm, it still was not ready.

An hour later, they decided to stop for the day and return the next morning. They planned to use a machine known as a *helicopter,* which has a large, horizontal propeller. It allows for faster and improved surface finishing.

When they came back the next morning, they tried the machine. Unfortunately, it didn't work. In fact, it made things worse, damaging the surface instead of improving it. After some time spent, considering options, we agreed that the only solution was to finish the slab by hand. It would take longer, but it eliminated the risk of further damage. It took forever, in reality at least several hours.

141

142

143

144

145

146

The entire time, I was deeply concerned that the concrete would not achieve its necessary strength: In New York City, every concrete pour is monitored by a special inspection agency. They usually collect four samples, cured in insulated boxes, called *curing boxes,* during the pour. These are then tested over time, once after one week, twice after four weeks, and once more after eight weeks, using a compressive strength machine. The results must meet the structural engineer's specifications, measured in *PSI (pounds per square inch).* If the required strength is not reached within eight weeks, the engineer can authorize acceptance if the results are close enough. But if they are not, the consequences are very serious.

Imagine having to demolish thirteen cubic yards of hardened concrete—about 350 cubic feet. It would take several days of mechanical removal. Worse, the embedded radiant heating system would almost certainly be destroyed in the process. Reinstalling that system and replacing the concrete would result in significant financial damage. Nobody wants to be responsible for that.

Fortunately, the concrete turned out to be strong enough. But I kept imagining what the conversation with the contractor would have been like if it had failed. No one wants to tear out work that is already complete. No one wants to admit mistakes, especially when correcting them would be extremely costly. Who is paying for this?

Looking back, I learned a valuable lesson from the experience: I will avoid pouring concrete in cold weather whenever possible. I would not have understood any of this without going through it myself. These facts, among the many unspoken truths of the trade, should be familiar to any capable architect. More importantly, this experience revealed how easy it is to assume a perfect world when working in drawings. In reality, perfection is nearly impossible. The more you learn about construction *means and methods*, the more you come to understand just how difficult it is to achieve them. And yet, perhaps only a trained eye can see those imperfections. As designers, we must carry that awareness into the planning process and stay connected to those who take on the challenge of turning a vision, once only imagined on paper, into built form.

The masons who built the two demising walls worked like a well-oiled machine. They arrived each morning with a team of six and got to work immediately. Two were stationed at the mortar mixing area, two handled the supply chain, carrying mortar bags and *CMUs* (concrete masonry units), and the final two set the blocks to

147

148

149

construct the walls. Their speed was impressive, and I found their efficiency deeply satisfying.

One of the walls, located adjacent to the long staircase, presented a particular challenge. Initially, I thought we would need extensive scaffolding inside the stairwell to complete it. But this raised serious concerns. The stairway was the only means of egress, so blocking it was not an option. Also, scaffolding rentals are typically expensive. We resolved the issue by hiring the same masons who were already working on the neighboring site. I asked the neighboring developer if we could build my west demising wall from his side, before his adjacent wall was constructed. He graciously agreed. Using the same crew for both projects had additional benefits, especially regarding material deliveries and logistics.

However, whether due to the unusual setup or simple carelessness, I began noticing that the quality of the masonry was not very precise. Tensions between the workers grew, especially between the two responsible for laying the blocks. They started arguing more frequently.

The boss was rarely on site. He typically showed up only for key moments—such as deliveries or *end-of-week* payments. During one of those visits, I shared my concerns. The horizontal bed joints on the second floor were completely off. One of the walls was nearly two inches higher on one side than on the other. The wall next to the stairwell was also out of plumb at a certain point. He tried to reassure me, claiming that the discrepancies would be corrected within the next two floors. But I had done my research. Codes and standards, such as TMS 402/602, the Building Code Requirements for Masonry Structures[14], clearly define the allowable tolerances for this type of construction. I reminded him of those requirements.

They did try harder after that complaint. The crew was changed, and new workers showed up. But the infighting started again soon after, and it became clear that the work would likely no longer meet acceptable tolerances.

At that point, the only tool I had was financial: I slowed down the *end-of-week* payments and documented the deficiencies. But what do you do when those deficiencies cannot be fixed?

The masons didn't return the following Monday. Nor did they come back after that. I ended up hiring another mason to finish the last few courses. In areas where work was delayed due to the setting of a steel column, I completed the masonry edge myself. It was certainly not as refined as a professional's work, but it was done ... and I became a mason. Of course, this was not the outcome I had hoped for, though I did gain my first hands-on experience. More important were the broader lessons. Who is ultimately responsible for quality control?

150

In my role on site, I was responsible for supervision. But I made the mistake of trusting the workers too much, assuming they knew what they were doing. By the time I realized the quality was not up to standard, it was nearly too late. While I share some responsibility, it would have been difficult to guide unfamiliar workers who were more focused on speed (to satisfy their boss) than on delivering precise work.

That is not to say all of them were careless. I got along very well with one of the two foremen. But I realized that the pressure to move quickly may have also contributed to the conflict between them. What they truly needed was clear direction and oversight from their supervisor. While the boss did try at times, his limited presence made meaningful improvement unlikely. He could have recognized the problems during his site visits, but either he did not notice them, or he chose to ignore them, in favor of finishing the job quickly. I have seen this pattern before. Contractors, especially those in the later stages of their careers, often lose touch with the physical work their teams are producing. Part of the reason may be the increasing demand of managing logistics. But another part may be emotional: years of stress in a challenging profession can cause a kind of desensitization.

151

I made a similar observation while working with the steel installer, who remains a friend of mine. Over the years, though, I noticed how the stressful nature of the job began to wear him down. When building inspectors came by, for instance, he would sometimes disappear for hours, unable to manage the stress. Still, what remained remarkable was his passion for the installation. He loved being on site and took pride in showing his knowledge of how to install heavy steel beams – some weighing easily 500 pounds or more – using the simplest mechanisms. He was highly experienced and clearly put significant thought into his *means and methods* ahead of time.

I was fascinated by how accurately he could predict the way beams and columns would behave in the air once lifted, simply by tying them in a particular way and guiding them with ease using *lead lines*. When his predictions were correct, he would smile. We could have made more consistent progress if he had the manpower to work on the project every day. However, since he was also responsible for

152

153

154

155

steel installations at a neighboring construction site, and because different workers often rotated in and out, consistency suffered.

This contrast highlighted two organizational models. The master mason delegated fabrication to his crew with minimal oversight. He delegated significant responsibility to his foremen. The steel installer, by contrast, wanted total control because he loved that part of the job. However, his workers, hoping for more responsibility, saw little opportunity to prove themselves. That's why none of them stayed for very long. It is a difficult balance.

I've come to realize just how many different personalities exist among builders. While the steel installer could be described as an *acrobat* due to his refined *techniques*, others might fall into categories like *musclemen*, *detailists, piece-rate workers, or freelancers.* I'll refrain from defining these too precisely, but it's clear that people with very different mindsets and backgrounds come together on a job site. If you want to manage a project effectively and achieve strong results, you have to understand these different characters and guide them accordingly. I can't say which personality type I connect with best, but I know that building trust with everyone on site is essential for leading a complex project with varied goals.

I also have not yet said much about the workers responsible for the finishes – arguably the most important to the architect, since their work shapes the building's *craft* and final appearance. And, it relates most to the original drawings. I also have to admit that I enjoy my conversations with finishers the most. There seems to be a special connection there. If you are detail-oriented, you naturally think about *installation sequence* and *means of execution*. This focus on *technique* lies at the heart of both *design-build* architecture and construction. Some of the most memorable conversations I have had were with highly skilled craftsmen explaining their exact methods for achieving the best possible results. Since I began working on *design-build* projects (starting when I was probably sixteen) I have always thought of building as an essential part of architecture.

In this particular project, I built many components myself, sometimes with the help and support of my students who were eager and curious to understand this hands-on perspective. The most important contributions I made were the installation of the window/glazing system for the entire building, portions of the siding, all the kitchens, much of the flooring, and especially the staircase. I view the staircase as the architectural centerpiece of the project.

156

157

158

159

160

161

162

163

164

One of the most critical tasks was the installation of the glazing system. A building is often said to be *buttoned-up* when it becomes water-tight – a major milestone, since water infiltration can cause long-term damage. In this case, I had done extensive research, not just in selecting the right products, but also by hands-on training. I had already been involved in the *design-build* installations of similar *curtainwall* systems on two other projects. That experience gave me the knowledge and the confidence to combine the components efficiently. So I decided to install the glazing and window system myself, following the detailed drawings I had previously developed.

165

The system was complex. It integrated the structural window frame, fixed glazing with *curtainwall* technology, and operable windows into a single architectural element. Gasketing, in particular, became a learning experience. It is essential for waterproofing, but ultimately invisible – a hidden yet crucial part of the system.

While working on the glazing, which took several months, I was also managing the interior construction. The interior construction involved three main trades: light-gauge framing, plumbing, and electrical work. For the wall framing, I laid out the entire floor plan using permanent markers directly on the concrete slab. We had to be extremely careful installing these walls. Normally, framers nail the base channel into the floor with powder-actuated nail guns, but that wasn't allowed here due to the embedded radiant heating pipes. The crew didn't appreciate this complication, because again, the alternative approach took longer.

166

The contractor, who was Chinese, was smart, and well-organized. Having grown up in the U.S., he spoke fluent English. His workers, with one exception, did not. At first, they sometimes acted as if they couldn't understand my instructions, or may have been pretending not to understand them. But once they realized how frequently I appeared on-site, often unexpectedly, and how closely I monitored their work, they began to ask questions and they started to build trust. These exchanges happened mostly through gestures and demonstrations with tools. Surprisingly, this approach worked. It seemed we could communicate effectively through shared action. The discipline, in a way, had its own *language*.

167

Often after these discussions, I would start doing certain tasks myself. My requests sometimes meant additional work, and naturally, workers are reluctant to take on more than they are paid for. *Time is money.* But when I took the pressure off and handled some tasks myself, things moved more smoothly. We were able to progress and reach the results we needed. A more persistent challenge was the consequence of my own frugality. I had made it clear to the framer, plumber, and electrician that this was a low-budget project. During the bidding phase, I pushed hard to get the prices down. Once a contract is awarded, you lose leverage, so you have to be strategic beforehand.

Unfortunately, subcontractors who bid low, often overbook themselves to ensure a steady stream of work. They may plan from the beginning to not appear regularly at your jobsite. Initially, their attendance at my construction site was generally acceptable. But later, they appeared just two or three days a week. Although I paid them on time and we made progress, it was not always at the pace one might expect. Whether it was truly sufficient is, of course, a matter of opinion. Their commitment to my project was not based on my site conditions, but rather on how things were going at other jobs – especially those, where the general contractors had not paid them or had not treated them well.

In the end, it's debatable whether I actually lost money by moving slowly. While waiting for various subcontractors, I worked on parts of the project that had to be done anyway. Some days were very slow. But since I did that work myself, I also saved a considerable amount of money. That included much of the façade and glazing, kitchen installations, the staircase, flooring, and numerous small but significant details like baseboards, drywall edges, invisible door headers and other finish components that will tease the eye. These small *touches* are always at the detail scale, they are very time-consuming, and they are the ones that shape our everyday experience of space and craft.

A developer would argue that I lost a lot of money. If I had pushed for faster completion, the apartment units could have been rented out much earlier, and the asset would have started generating income sooner. I can see that point, but I am not sure where the *sweet-spot* lies in that equation. My perspective was quite different from that of an *economic* developer. I believed that by having the time to complete the self-built portions of the project, I didn't just save money, I also gained more control over precision and appearance of significant architectural features. I even had the opportunity to revise certain details on the spot. If I had tried to do that through a contractor, it would have led to delays and cost increases through formal change orders. These legal instruments often spark disputes, and they can become very unpleasant and time-consuming.

168

169

170

171

172

173

174

175

176

177

178

179

When aspects of the design are not fully resolved in the planning phase, contractors quickly notice the gaps – and sometimes take advantage of them. In this project, I had zero change orders. I avoided that whole issue by agreeing to do certain parts myself. In that way, waiting periods became productive, often turning into trouble-shooting sessions. Without those necessary *take-a-breathers*, I wouldn't have had time to think and the project wouldn't have moved forward, at least not in a controlled way.

That process also brought me closer to the on-site community. It nurtured a deeper understanding of the project. I was patient and I didn't mind. I wanted to be on the construction site as often as I could. The workers became increasingly curious about what I was doing. They had not experienced anything like this before. Over time, they started helping me and offering useful, practical advice.

But also, I realized – having been exposed during my architectural education to buildings constructed centuries ago, and to the intense attention given to detail in the pursuit of architectural beauty – that our era seems to be in trouble: It can't be in the interest of our culture to prioritize finishing quickly over the producing work of substance and sophistication. If we keep heading in that direction, we risk losing our culture entirely. Architectural production would become nothing more than a commercial activity, and only serve the purpose of running a profitable business. I am not opposed to making a profit. But I am certain that architecture, as a discipline, cannot be defined solely in those terms.

We are responsible for giving something meaningful and inspiring to the experience of our existence. We need this in order to be human. If this purpose is lost, architecture will be lost as well. The way architecture is currently *processed* often serves only the logic of profit. If we want to change how architecture is understood (if we want it to mean more than helping an industry maximize profits), then we will have to resist that trend. Otherwise, our work will fade into irrelevance. The kind of sophisticated thinking and traction that architecture requires doesn't fit easily into how money is made today. Do we have to fit? Or, what exactly should our niche be? We should practice our own *language* and take advantage of the current conditions as we go. I believe that most people – at least those not damaged by the economic hardships they've faced – will still be inspired by the way architecture is producing something inspiring and tangible. Something that can be felt and experienced, forming memories and events.

180

181

182

Earlier, I said I didn't favor one philosophy of craft over another. But that is not entirely true. I favor those who are focused on their finishes and on the architectural beauty that results. I once learned from a craftsman that the essence of a good outcome lies in preparation: That the anticipation of what is essential in fabrication must be deeply considered before any action begins. Preparation wins the game. And this is very similar to our profession. We thrive by predicting conditions, by training ourselves to see what matters in the execution of a complex system. One could argue that a finishing craftsperson does the same, only on a smaller, more intimate scale. There is often very deep knowledge in that kind of work, paired with pride and a passion for achieving something beautiful.

So why not take advantage of the craftsperson's knowledge, competence and pride in their work? We should be able to learn from them. We should learn to work *with* them. We know, from experience, that they are approachable. We know they are often eager to do something special, when it presents a challenge to them that they appreciate. But the way our profession is currently structured (look again to the *AIA A201* [9]) prevents any real overlap in activities. That makes collaboration nearly impossible. We have the resources. We have the materials. We have the *know-how* of people who care about building something meaningful – something they will be proud of. Why aren't these resources connecting the way they should, the way they did centuries ago? As architects, we must reconnect with the material world. It should be our main target. The craftsperson is the gatekeeper to that world, the one who makes it possible.

183

184

185

186

187

188

189

190

191

192

193

194

195

196

197

THE LAW

198

199

200

For a discipline that creates spaces used by people every day, certain legal constraints are inevitable. The architect must be well-versed in the rules and regulations that govern the profession. The most important of these are compiled in building codes. By my count, the *New York City Building Code* alone is more than 900 pages long. In addition, there are separate construction codes for plumbing, mechanical systems, energy conservation, there are zoning regulations and accessibility standards and many other areas that we, as architects, should be familiar with. How can anyone possibly remember all of this and know which codes apply when? This is even more confusing, since the *body of law* is always dynamic. Things change that you might not be aware of. For instance, there are *local laws* that can be passed quickly, for instance related to *bird-friendly glass*, in order to prevent bird calamities in New York, based on findings that the reflective curtain wall environments lead to hundreds of thousands of bird crashes and deaths. After that local law is passed, you must be knowledgeable about it, because it might affect your specifications.

In New York, there is a profession known as the *expeditor,* essentially a professional code consultant. The best expeditors do exactly that: they consult on and help interpret codes and regulations. Others focus more on managing the interface with the Department of Buildings *(DOB),* the Department of Transportation *(DOT),* The Department of Environmental Protection *(DEP)* and many more. They are handling forms and dealing directly with various departments and plan examiners. Sometimes these transactions are very tedious.

For architects, it is not enough to know the codes; you must know which specific forms to file and how to complete them accurately, checking the right boxes to ensure progress. Administrative procedures exist that you are unlikely to be aware of when first entering the profession. Because of such complexities it is almost obligatory to hire an expeditor, if you don't want to get into trouble, but also in order to have a *teacher* to learn for the next job. Even with this knowledge, progress is almost always slower than expected. Weeks often pass before you receive an update from the Department of Buildings. My impression is that this delay is due to understaffing, shifting personnel assignments, and other bureaucratic factors. Very

201

202

203

often, though no official statistics seem to exist, your first request will be *rejected*. Often the feeling is that the examiner arbitrarily places boilerplate objections on the drawings (even if they do not apply, or the objection is unfounded) during the first review to more quickly move the filing off their desk, and back to the architect and expeditor for revisions. It would be helpful to have a personal connection within the Department of Buildings, so that practical, common-sense decisions could be reached more quickly and collaboratively. However, the department does not seem to be inclined to engage in direct, informal exchanges. Perhaps this reluctance stems from a desire to avoid overly close relationships that could lead to corruption, or perhaps it is simply a way to sidestep legal responsibility for decisions made in complex approval processes.

During my time as an architect in Berlin, I experienced almost the opposite approach. When a conflict over code interpretation arose, the examiner from the building department took an active role in finding a solution. She called a meeting with all stakeholders, offered thoughtful suggestions, and helped reach an agreement that everyone could support. In New York City, such a collaborative approach is rare. Here, as soon as a conflict emerges, the best chance of progress is often having an expeditor to negotiate on your behalf. In this setup, the architect plays the most disadvantaged role. Most of the required forms are signed and sealed by the architect, which on paper makes them responsible for any problems. Yet, despite carrying significant liability, the architect has seldom a direct voice in the negotiation process. A similar dynamic plays out on the construction site when inspectors visit. Many issues are quickly blamed on the architect, with the explanation that they failed to exercise the proper documents. But often it is not the proper documentation that causes conflict, but an underdeveloped connection between such documentation and the oversight at the construction site. Within the building department, there are two distinct roles. One is the *examiner*, who works entirely in an office environment. The other is the *inspector*, who specializes in visiting construction sites. Their mindsets and methods are quite different. The *examiner* is typically cautious, raising *objections* early in the review phase to address any perceived concerns. Even if you, as the architect, believe your proposal is fully compliant, the burden of proof falls squarely

204

205

206

on you. This means being deeply familiar with the hundreds of pages of building code, ready to cite specific sections to defend your position. Here, teamwork with proficient code consultants is essential, because they know more than you.

Given the complexity, it is nearly impossible for an architect to be fluent in every provision of the code. Conceptual problem-solving in architecture rarely involves the repetition of fixed, factual answers, so the breadth of knowledge required can feel overwhelming. This is precisely where a skilled code consultant becomes invaluable – their daily work exposes them to these disputes regularly, giving them the experience and insight to navigate such challenges efficiently. The *examiner's* role involves navigating many complexities, but their ultimate responsibility is limited. Once all objections have been addressed, you will eventually get permission to begin construction. The approved drawings will carry a stamp on every page: *"Examined for zoning, egress, and fire prevention only, as per directive 2/75."* [15] This means that even after a lengthy process of adjustments, negotiations, and reviews with the *examiner*, you are still not fully in the clear. There is always the possibility that something – outside of *zoning, egress,* or *fire prevention* – was overlooked. If that happens, you remain liable, even if construction is well underway or already completed. The directive itself states, *"The procedure takes cognizance of the professional standing of architects and engineers and their responsibility and capability to provide plans of buildings, which are in compliance with law."* [15] In other words, the burden rests almost entirely on the architects (and engineers). The *examiner* is shielded from responsibility, even though they encounter similar issues (and thus knowledge) far more frequently over the course of their career. One could argue that the *examiner's* responsibility and capability should at least match that of the architect, if not exceed it. Yet the system appears designed to protect the *examiner* and the Department of Buildings from liability rather than fostering a culture of collaboration and shared accountability. This dynamic creates a climate in which each party looks out primarily for themselves, rather than focusing on an outcome that reflects a shared civic pride. Too often, there is a lingering suspicion that the process is driven solely by profit, and that believing otherwise is naïve. But as architects, we are motivated by a genuine concern for our environment. Are we alone in that?

207

208

209

210

211

212

Interaction with an *inspector* is an entirely different experience. *Inspectors* tend to focus on safety conditions and on whether the built work matches the approved drawings. Their scope is narrower than that of an *examiner*, and some appear to have favorite codes and orders they check repeatedly. The result is a predictable pattern of citations. And those citations are no small matter – *inspectors* issue many summonses, often with steep penalties. A *Class 1* summons, for example, typically ranges from $2,500 to $5,000. This is not a parking ticket: on a *bad day*, multiple summonses can cost a small fortune. And, the bad day is fully dependent on the *inspector's* mindset and mood. An *inspector* will often try to issue as many violations as possible on a single visit, not just one. While *examiners* tend to interact repeatedly with the same *expeditors* (people who understand the rhythm and temperament of their work), some *inspectors* on site, tasked with construction safety enforcement, typically engage with a construction crew only once. I have never had such an *inspector* return several times to the same project. They could, but they usually do not. Instead, they tend to make their visit count. Weeks or months later, a different *inspector* will appear.

Inspectors hold a powerful position: they can issue tickets for any code violation they observe or interpret. Their knowledge of these violations will almost always exceed yours, because they deal with them daily. It is difficult to challenge their expertise; they are always a step ahead.

There are two different types of summonses they can issue. *Class 1* summonses, as mentioned, carry a substantial monetary fine. *Class 2* summonses are considered "curable," meaning you can correct the problem and submit photographic evidence of the fix. Regardless of the classification, *violations* come with a hidden cost: while the matter is under review, a *stop-work order* is placed on the site. This means construction halts entirely until the issue is resolved.

The penalty, therefore, is twofold. First, there is the financial hit. Second, and often more damaging, is the loss of valuable time. A site that has been progressing smoothly can all the sudden be thrown significantly off schedule. The waiting period to have the *stop-work order* lifted can stretch into several weeks, sometimes even months. Every cited item must be corrected to a faultless standard. Only then can you request

213

214

215

a *re-inspection*. The building department assigns a date, and an inspector returns to verify compliance. My *re-inspections* generally went well. Perhaps we did a good job. Perhaps the specific issue was never really the main concern. In any case, it is always a relief when the people who have the power to upend your schedule finally show leniency. But is this truly the only way the system can operate? I find myself wishing for more collaboration, for more common support for our architectural culture.

Another area of vulnerability is less obvious at the start of a project. Early on, you are required to put a standardized *Work in Progress* sign on the construction fence which displays mandated information about the project. The information includes the owner's address. In practice, this means that if anyone wants to cause trouble, you are exposed – potentially without realizing it for a long time.

In New York, and likely in many other parts of the United States, there are people who make a habit of exploiting these signage requirements in order to create problems for a particular gain. Sometimes their motives are personal or financial; sometimes they stem from unrelated frustrations or wanting to stall or stop construction progress. The owner's name on the sign is typically that of a *Limited Liability Company*, or *LLC*. If someone files a lawsuit against the *LLC*, the owner is immediately at a disadvantage. The law dictates that an *LLC* cannot represent itself. It must be represented by an attorney. This costs money. It creates an opening for an entire *legal industry* to exploit the situation, knowing that you will have to communicate through costly legal counsel. To me, this is an injustice. The law often does not follow the facts; it follows the procedures. These procedures can drag on for years, costing thousands of dollars, all while you defend yourself against someone who knows how to exploit the system.

On the day construction began, I did not fully realize what I was stepping into with respect to *inspector* visits. I felt confident that my drawings had been approved for construction, and thought this would go a long way. Yet only a few weeks into excavation, I arrived one morning to find my workers uneasy and nervous: the *inspector* had just shown up. He was already inside the site *inspecting,* and it was immediately clear from his *objections* that we were going to have a hard time with him.

The *inspector* was confused at first because excavation was being shared between two adjacent lots. Both lots had hired the same excavation contractor, which made perfect sense logistically but complicated the *inspector's* perspective. Who, he wondered, should receive the summons? In the end, he shut down both job sites, citing safety concerns about the protected edges and the required *angle of repose* of the excavation. He demanded that we file a document with the Department of Buildings showing all anticipated phases of excavation, even though separate drawings for each site already existed. It was within his right to request whatever he believed contributed to safety, but we knew that meeting his demand would take time. Creating the drawings and having them processed for approval would inevitably delay us significantly.

Before raising the excavation issue, he complained about something *inspectors* always target: cleanliness on the site. He objected to what is generally called *housekeeping.* On an active construction site, keeping things clean is a constant struggle, especially as more workers and trades are added. His argument was dramatic. He said that if someone were to have Sunday breakfast on the balcony of the adjacent building, they "*would see all of this shit lying around.*" His crescendo of complaints reminded me of my year in military service, or my early primary school days under teachers who were former Second World War veterans. It was clear that his main objective was not simply to enforce rules but to establish authority through confrontation.

As expected, we received a full *stop-work order*. The site was shut down until further notice. We had to produce statements from the *SOE (support of excavation) engineers* to clarify the issue of the two adjacent lots. We also had to submit photos to demonstrate improved cleanliness. The *clean-up* summons was classified as *Class 2*, meaning it could be cured, but the excavation summons came with a substantial fine. Worse, it was issued twice – one for each lot – so the penalty was doubled, to the city's advantage.

216

217

218

219

220

221

While I was devastated, my neighbor remained calm. He viewed the incident as just another business expense to be anticipated. The process of resolving the issue took more than three weeks, and the fines were costly. Both the financial setback and the delay weighed heavily on me and made me anxious about the challenges still to come. The second significant visit from an *inspector*, about two years later, was even worse.

One of my workers told me that an *inspector* was on site. He had slipped in through the entrance door during the late lunch break, when workers tend to leave the door open or propped, and had gone straight to my neighbor's construction site. Since we had shared excavation logistics earlier, we were also sharing a continuous construction fence. This made us more vulnerable: more workers passed through the door, and it was often left ajar for easier access. *Inspectors* know these loopholes. They sometimes spend hours watching from across the street before making their entrance.

When I was told the *inspector* was around, I searched everywhere, up and down the stairs, into the garden, but he was nowhere to be found. Then it dawned on me. I checked the outside of the fence and found a *stop-work order* stapled to the plywood. I immediately called the building department at the number listed on the document and asked them to contact the *inspector* who had posted it, but nothing came of it. There was no chance for any interaction.

That day we accumulated six tickets. The *inspector* returned the next day to hand them to us in person, smiling broadly as he did. I asked where I could have found him at the time he claimed to be on site. He said he had been there. In small talk, he proudly told us that he writes about 800 summonses a year. Multiplied by the typical fines, that amounts to millions of dollars in revenue for the city. From my perspective, it seemed clear that after posting the *stop-work order* around 3:00 pm (an hour before the end of his shift), he had left the site and written the tickets later, perhaps on his way home. I cannot prove it, but I suspect his phone records and city-issued computer could show his location at that time. But this information was not available. My own cellphone records prove I was on site, since I called the department during that exact period.

The tickets totaled $20,000. One was for not having a *competent person* on site – a person with certified safety training. That person was me. Again, I had evidence showing I was present.

I contested the summons through *OATH*, the *Office of Administrative Trials and Hearings*, which allows for *objections*. An online hearing was scheduled. The judge, an employee of the same city system, sided with the *inspector's* claim that he was on site and that it was impossible for us to have missed one another on such a small job. Even though my evidence clearly placed me there, I lost. All six summonses were enforced, and I had to pay. It was the saddest day of the entire construction process.

222

Looking back, was the outcome predictable? I think so. If the judge and the inspector work for the same system, the chances of success for the architect are slim. Once again, everyone seemed left to fend for themselves, except the city. There must be better ways to collaborate.

One of my last encounters with an *inspector* on site was much better and it restored some of my trust in the agency. On that day, I realized that some *inspectors* actually do have empathy. They seemed genuinely interested in communication and in working with us to improve safety conditions. He arrived together with two other *inspectors*. The sight of three *inspectors* knocking on the door first thing in the morning almost gave me a heart attack. Two of them were from the *Energy Compliance Unit* and had come together. The third, unaware of their visit, told me to take care of them first and said that in the meantime he would walk the building on his own before speaking with me. That situation (, which experienced contractors warned me about,) made me nervous, since I knew he would see everything without any explanation or context from me.

223

After the *Energy inspectors* finished their review (requiring me to file an amendment to the drawings on record), I joined the other *inspector*, who by then had already completed his walkthrough. This was an obvious disadvantage, since I had not been able to accompany him. He greeted me kindly, even joking that I might want to see his ID for the daily log. Of course, I was required to enter it, but in my nervousness, I had not even thought to ask. He seemed to notice my unease. Instead of escalating the situation, he calmly shared his observations. He said that overall, I had done a good job as *Construction Superintendent*. The site looked clean and organized. Safety guards were installed at the right heights and in convincing ways. Materials were stored thoughtfully without interfering with circulation paths. He had also reviewed the log books on site. They were complete and thoughtful.

224

225

226

227

228

229

230

231

232

233

234

235

236

Maintaining those records was no small task. Every day required extensive documentation. The most important were the *Construction Superintendent's* log and the *Pre-Task* log. The *Pre-task* log is particularly challenging. At the start of each shift, the *Construction Superintendent* is legally required to call a meeting – or several, depending on the crews present – with the workers on site. Together they review the scope of work for the day, and the safety concerns connected to that work. Specific responsibilities must also be assigned, including that of a *competent person.*

A *competent person* oversees the task and monitors compliance with safety standards throughout the process. Typically, this is the most experienced worker. He or she must also have advanced site safety training, developed jointly by the city and *OSHA* to promote better education and reduce accidents. To qualify, the *competent person* must complete 62 hours of certified training, which must then be renewed every five years through additional courses.

The problem is that many workers often see the *Pre-task* meeting as a waste of time. They want to get started right away and feel that they already know best how to do their jobs. As the *Construction Superintendent*, however, it was my responsibility to enforce these meetings and to ensure that they were documented every single day, with the signatures of all workers present. I also had to keep proof on file that every worker had completed at least 40 hours of certified safety training. The *inspector* was impressed to find all of the log books up to date and additional records of all 40-hour safety training certificates. I felt a wave of relief, but only for a moment. Then he told me that he had identified one major *objection*. My confidence vanished and I began to sweat again. He led me into the stairwell, looked up, and asked: *"What do you see here that should concern you?"* It was a difficult question, where you could reveal more than you'd like. Keeping a job site safe is a complex responsibility, and there were several possibilities. I had two guesses about what he might be referring to. I chose the more obvious one: on the upper run of the stair, I had already closed the risers so that nothing could fall through and hit someone working below. But on the two lower runs I had not yet completed this work. I admitted to him that I recognized the issue and was in the process of addressing it. He then asked when I planned to finish. I explained that I was also a professor at *City College of New York*, where I was expected to teach in about an hour. Since no workers were on site that day, I had not planned to complete the correction immediately. I asked

237

238

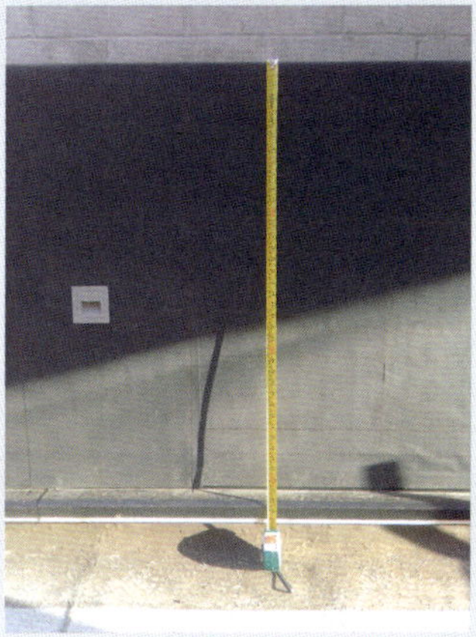

239

if I could finish by the following morning. He raised his eyebrows and asked if I really wanted to leave the construction site with such a serious safety issue unresolved. Realizing what the right answer was, I quickly replied, *"Of course not. I will make the necessary adjustments immediately."* That was the response he was looking for.

His point was valid. On any construction site there is always too much to do, and it is easy to prioritize tasks that push the project toward completion over those that *only* improve safety. The temptation is strong to delay the latter, because finishing visible work makes you feel accomplished at the end of the day. But postponing safety measures is dangerous and shortsighted. Being reminded of this by the inspector was important. I explained how I planned to correct the issue. He reminded me that the responsibility was entirely mine and that inspectors are not supposed to give advice. Still, he shared how he had seen the problem solved on other job sites. That was both helpful and unexpected.

For me, this meeting was eye-opening. I had never before experienced a conversation with a New York City *building inspector* that was memorable in terms of genuine human interaction. This *inspector* showed that it is possible to work together and find common ground. He cared, and he showed empathy. For once, I saw in him a *role model* of what a proficient and sympathetic *inspector* could be.

I corrected the riser problem that same day. As the *inspector* left, he told me he would return for a *reinspection* the following morning. I actually looked forward to seeing him again, but he never came back – likely because he knew I would follow through on what we had discussed. Perhaps that was his sense of humor, perhaps it was his way to be efficient. Still, I wish I had been able to thank him for his empathy and for the way he interpreted his job.

Another *inspector* story stands out for me, because it taught the most important lesson about the architect's reputation on a construction site, at least through the eyes of many *inspectors*. It also gave me guidance on how to run a project more smoothly next time. Many *inspectors* are primarily concerned with the consistency of built conditions in relation to the approved drawings. That sounds straightforward, but it is especially challenging when it comes to plumbing installations.

240

241

242

243

244

245

246

247

248

The same *plumbing inspector* came to most of my inspections, and over time we developed quite a rapport. *Plumbing inspectors* are usually assigned to specific zones in the city, and the one in my area never missed a date. The problem was this: plumbers often rely on their own sense of *standard practice*, built over many years of experience, rather than double-checking the specific pipe diameters and details in the drawings. From my position as *Construction Manager*, it was difficult to recognize on the spot whether a pipe diameter was half an inch off. I should have done a better job. Whenever the plumbing inspections took place, particularly the rough-in inspections before walls could be closed, the *inspector* frequently spotted inconsistencies. This was problematic for two reasons. First, he always blamed the architect for the errors. *"Tell the architect he should really check the drawings",* he would say, even though the responsibility more often rested with the plumber or the mechanical engineer. I decided not to reveal who the architect actually was. Instead, I politely suggested that in a business this complex, no single person should be held accountable for all mistakes. The *inspector* looked at me, puzzled, perhaps because it was not the kind of comment he was used to hearing on a construction site.

From my time as *Construction Superintendent*, I had learned that workers often see the architect as the default scapegoat for anything that goes wrong. This is ironic, because according to the *AIA's* standard construction contract *(A201),* the architect is explicitly not responsible for construction *means and methods*. How then can the architect be blamed for errors that surface precisely during that process? Perhaps the profession's effort to waive responsibility for on-site operations has backfired. Many parties in construction still identify the architect as the source of error. It is easy to blame someone who is not usually present on site.

Still, the lesson is clear: the architect will often be the subject of blame, particularly when miscommunication arises. And the blame can come from multiple directions. After an inspection, a plumber may refuse to redo his installation, even if it does not follow the engineer's drawings. From his perspective, tearing out work already installed would cause greater disruption, such as damaging wall framing, which would slow down progress even further. More often than not, the architect ends up coordinating *amended* drawings with the mechanical engineer and the expeditor. Developers typically prefer to pay for additional paperwork rather than undo completed work.

The best approach, of course, is to avoid *objections* altogether. They create endless headaches, particularly for the architect, and they cause significant time delays since *examiners* must approve *amendments* after drawings are revised. The *examiner's* environment within the Department of Buildings is quite different from the construction site. It is calmer and follows its own deliberate pace. At the beginning, when filing the first set of drawings, the *examiner* took his role very seriously and returned many pages of *objections* and requests for clarification.

Bit by bit, these issues were addressed and the drawings revised. Yet sometimes this process led to conflict, especially when the architectural intention was at risk. That was always the most critical point. Unfortunately, it is difficult to communicate such concerns, since no one but the architect tends to view architectural objectives as strong enough to justify altering established viewpoints.

I remember when Julius Posener, the noted modern architecture historian in Germany, gave a lecture at my first university, the Technical University of Berlin[16]. In an informal talk after the lecture, he suggested that architecture should be taught in primary schools, so that children would grow up understanding how important design is for shaping our society and culture. This would influence the thinking of everyone, from *examiner* to *inspector*. We are still far from achieving that vision. Building codes, of course, are not about architectural intention. They require careful adjustment and negotiation. The *permissible* and the *artistic languages* must be coordinated with one another. In my experience, this worked best when I was able to attend scheduled meetings between my *examiner* and *expeditor*. Normally, architects do not participate in these meetings, but I was able to obtain a badge that allowed me to enter the Department of Buildings and join the discussions.

The *examiner* was cautious, as expected. I later learned from a former student of mine, who had worked as an *examiner*, that they are instructed not to suggest solutions directly, out of fear of future liability. However, after some time, and after small talk on topics such as politics, the *examiner* became more approachable. If you asked the right questions and listened carefully, he would often hint at possible solutions without stating them outright. My *expeditor* understood this *language* well and knew how to adjust the design so that the concerns would be addressed while still remaining permissible.

249

250

251

252

253

254

255

256

257

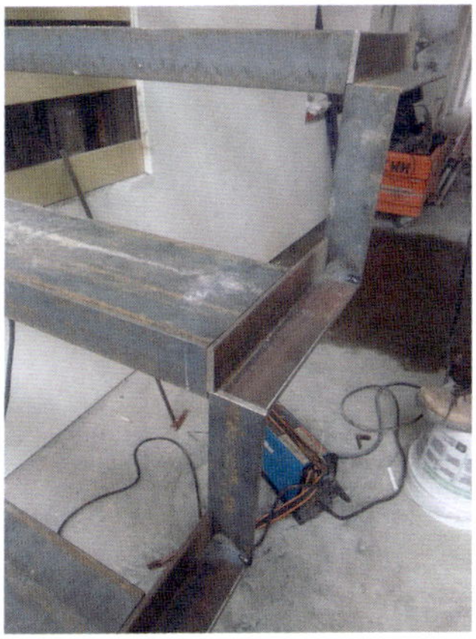

258

259

260

Not all discussions went smoothly, and sometimes conflicts could not be resolved. For example, I had planned a small interior space on the roof of the top apartment, connected to the lower floor by an interior stair. My argument was that this space was part of the fourth-floor apartment. The *examiner* disagreed. In his view, it constituted a fifth story, which meant that the building would then require an elevator. That would have placed the entire architectural layout in jeopardy. It was a situation where building codes and architectural intention could not be reconciled. In the end, I had to sacrifice the additional roof space and abandon my idea of how the roof could be used. Eventually I got the drawings approved and was ready to start construction. Yet the story about the additional space on the roof did not end there. Months later, while I was at the Department of Buildings filing some administrative paperwork, another *examiner* recognized me in the elevator and remembered my project. Since not many architects tend to stand out in that place, I took it as a small moment of recognition.

261

She told me that my roof design had actually been used as a precedent in a conference where *examiners* discussed recent *code changes*. She said that if I had filed my permit only a few months later, I might have been allowed to keep the additional roof space. I asked for more details, hoping there might be another chance in the future. But before she could explain further, the elevator doors opened. She quickly exited, saying only that it was good talking to me. I tried to stop her, but she was gone. At that moment I understood her caution. If *examiners* are truly not allowed to guide architects in shaping arguments, even when they feel sympathetic, then trust in communication will always remain limited. But does it really have to be that way?

262

Afterwards, I tried to piece together what she might have meant. I contacted my *expeditor* and several other colleagues who were knowledgeable about recent code updates. Finally, one of my sources, a former Department of Buildings employee who now works as an independent code consultant, figured it out. A directive had recently been issued, in between formal code updates, that rewarded buildings with green roofs.

263

Under this order, buildings with sufficient *Floor Area Ratio* (*FAR*, or the maximum permissible square footage) could add a greenhouse structure on top of a green roof. I cannot be sure if this is what the *examiner* was hinting at, but it seems likely. My project did have the *FAR* available, and I am certain that a developer would have pursued it. But by the time I learned of it, the process was already too far along, and I no longer took the possibility seriously.

264

This kind of knowledge is incredibly valuable, which makes the role of the *expeditor* in New York fascinating from both the developer's and the architect's perspective. Among the thousands of *expeditors*, some specialists know the mechanisms that can secure additional square footage, unlock material alternatives, or achieve better layouts. Returning to the point I made at the beginning of this chapter, there are simply too many codes for one person to memorize. The *expeditor's* expertise becomes essential for advancing a project efficiently while defending architectural objectives.

265

Yet this system also reveals a flaw. The reliance on *expeditors* is a direct result of the lack of open communication between the Department of Buildings and the architectural community. It reflects the department's strategy of avoiding responsibility for architectural decision-making. Many *expeditors* gain their expertise after working inside the department itself, where they accumulate knowledge that could be invaluable for better architecture. Instead, this expertise is redirected into private consulting businesses. The prevailing culture, once again, is best described as *everyone-against-everyone-else*, with architects left at the weakest point of connection.

The developer seems more connected to this world, particularly when it comes to the well-known reality that *time is money*. Much legislation is tied to that principle. *Property taxes*, which must be paid even during construction, are the city's primary tool for generating revenue. In New York, *property taxes* are extremely high.

266

The longer a project takes, the more expensive it becomes, not only because of those taxes but also due to the ongoing expenses of maintaining a construction site without generating any income. The city appears to soften this burden in theory. By law, property taxes cannot increase during the course of construction *(§ 11-209 Taxable Status of Building in Course of Construction*[17]*)*. In practice, however, the Finance Department does not seem to follow its own legislation. Each year during construction, my property taxes were raised substantially. To protect myself, I hired a real-estate lawyer to file a petition with the city to correct what I assumed was a clerical error. For three years we filed and refiled petitions, but there was never any

267

268

269

acknowledgement, response, or recognition from the city. The silence seemed intentional. By ignoring their own legislation, the city generated more money. The lack of communication worked in their favor. The only updates I ever received came from my lawyer, reminding me to renew the petition each year, for a fee. If I failed to renew, the claim would expire, and with it, any chance of receiving a corrected tax evaluation. For three years, I kept paying the lawyer, believing the relatively small legal fees might eventually save me substantial money once the Department of Finance responded. But when the building was nearly complete and there was still no reaction, I gave up. There seemed to be no realistic way to strategize, or perhaps no way at all. Whether intentional or not, the fault was in the system. Developers did not speak about this, it was simply part of doing business. For me, if there are rules, then everyone should follow them, especially the very institution that created them. Perhaps I am too pessimistic, and maybe one day the petition will resurface and be reviewed. Is it likely?

This experience reflects a broader issue with clerical processing throughout the project. Submissions to various departments often took many weeks to be reviewed. More than once, after weeks of silence, I finally received a response only to be told the application was rejected because of a missing document. A simple first screening on the day of submission could have prevented the delay. Instead, the process penalized the applicant.

Why should it be this way? With constantly improving digital systems and the rise of *artificial intelligence*, repetitive procedures could easily be automated, eliminating much of the lag time. I often wondered whether the clerks handling these requests understood the true cost of waiting. A few weeks of delay, which was routine, could mean the loss of thousands of dollars. If there were more compassion and accountability, the culture surrounding architecture and construction would be more efficient, humane, and enjoyable. It is striking that for some stakeholders money is treated as the highest priority, while for others it seems entirely irrelevant.

The rest of this chapter I will dedicate to one of the most difficult legal problems I have ever faced, not only during construction but throughout my entire life. For a long time, I was not even aware of this particular vulnerability when constructing a building. I had created a *Limited Liability Company (LLC)* to protect my personal legal status. The entire construction operation was organized under that *LLC*. As required, the owner's information was posted on the large sign fixed to the construction fence. The *LLC* was therefore presented publicly.

One day, in the middle of construction, I received a *certified letter* informing me that I had been sued. The description of an alleged accident stretched more than thirty pages and was frightening: it claimed a person had been permanently and severely injured. Because an *LLC* cannot represent itself in court, I was required by law to hire a lawyer. This alone was deeply concerning. Litigation lawyers are notoriously expensive, charging hundreds of dollars per hour, and they are skilled at creating work that translates into billable hours. They often bill at 1/10 of an hour, every six minutes. They also know how to write contracts that give them maximum control – something architects rarely manage to do.

270

What had happened? An individual, was riding home on his electric scooter. He claimed to have been injured after striking a pothole in the public roadway in front of my property. Importantly, this was not on the sidewalk but on the city street itself. He sued the City of New York, but he also sued the adjacent property owner, which was unfortunately me.

271

In the City's preliminary hearing, called a *50-h transcript,* it was revealed that he had been riding against traffic on a one-way street, sometimes on the sidewalk, sometimes on the street, where the alleged accident occurred. He wore no helmet. It was 9:00 p.m., dark, and the headlights of oncoming vehicles likely impaired his vision. He claimed to have fallen because of a seven-inch-deep pothole. Yet, the pothole was never found. There were no photographs of the pothole in the original claim file, and no witnesses were named. Nevertheless, this plaintiff, who had been involved in similar lawsuits before, decided to pursue it.

So did his attorney's office, which had represented him in previous cases. They likely handle hundreds of lawsuits of this nature simultaneously. Their advantage lies in their mastery of *process*. They know the system far better than the defendants they sue. I learned that they typically work on *contingency*, meaning they take a percentage of whatever funds they recover, while the plaintiff pays no fees. Meanwhile, I questioned how the city itself would respond. Shouldn't the law ultimately serve justice and humanity, not just procedure? What I found instead was that the New York legal system takes these claims very

272

273

274

275

276

277

278

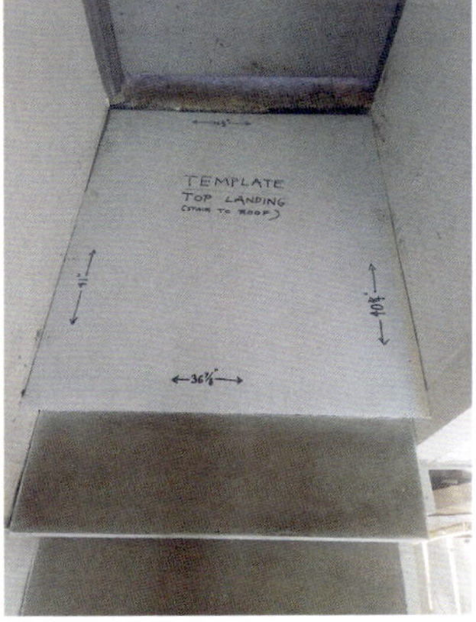

279

280

281

282

seriously and makes it extremely difficult to escape them, even when they appear dubious. My case has now dragged on for several years. In my opinion, the longer it lasts, the more it provides a steady income stream for all the lawyers involved.

Understanding the gravity of the situation, I decided to visit the plaintiff at his address and rang the doorbell. He came down in sports clothes, wearing very expensive sneakers. When I told him that I was there because of his accident, he looked puzzled and asked, *"What accident?"*

During our conversation (which you can probably imagine was awkward), I eventually asked him to dismiss me from the lawsuit. After all, the accident had not even happened on my property. He admitted that he wasn't really aware of who was being sued. It became clear that his lawyer's office was deciding whom to target – it was all part of a strategy. Then he asked what I would be willing to pay. When I hesitated, he grew irritated. I told him I would need to consult with friends who knew more about this sort of matter. His response was curt: *"Make it worthwhile."*

As I was leaving, he shifted the conversation. He asked if I had a job for his sister, perhaps as a *flag person* on the construction site, since she was unemployed. I explained that I didn't have any positions open. He gave me his phone number, asked for mine, and told me to call him. Less than an hour later, he called me himself to ask whether I had made a decision.

A week later, I called him back and offered a few thousand dollars if he would consider releasing me from the lawsuit. He refused, saying, *"We'll see each other in court,"* before hanging up. I have had many sleepless nights over this. All of this seemed so unfair.

I was raised to always reflect on whether I might be wrong or right in any conflict. In this case, it was clear to me that the plaintiff and his lawyer cared very little about that question. Their focus was only on what could be gained. Does the legal system itself care about truth and fairness? I am not sure.

Hiring a lawyer was its own challenge. Some firms demanded *down payments* in the lower five figures, and these fees only multiply as the case grows. Every party that had been sued faced the same problem: enormous legal expenses. Sometimes insurance would cover the costs, but often they refused.

In my case, the defendants included the *City of New York*, my *General Contractor*, and my *LLC*. The more parties involved, the higher the costs, since correspondence, filings, and later depositions all add up quickly.

When I began searching for an attorney, I realized how critical it was to find someone I could trust. To buy time, I tried contacting the plaintiff's lawyer directly, hoping to negotiate an extension. The receptionist took my messages but never returned them. After several days of calling, sometimes multiple times a day, the receptionist grew impatient and told me: *"We usually only talk to attorneys or insurance representatives. Just give us their contact information."*

The expectation was clear: as a defendant, and as an architect, I wasn't supposed to speak to them directly. Eventually, I was given the principal lawyer's email address and wrote to request an extension. The reply I received was the shortest email I have ever gotten: just one word, *"Yes."* I could only assume the brevity was due to the sheer number of lawsuits that office was handling.

Eventually, I hired my legal representative. His advice was in line with what the receptionist had already hinted at: focus on possible *insurance coverage*. But the insurances of the defendants had their own strategies. None of them accepted responsibility for a pothole in a public street.

There is even an organization called the *Big Apple Pothole and Sidewalk Protection Committee*, created by the *New York State Trial Lawyers Association* to account for roadway and sidewalk defects in the city. The reason is an administrative code *(§ 7-201)*, which bars lawsuits against New York City for such defects unless the city was notified at least 15 days prior to the accident. In my case, the city's lawyers argued that if there was a problem with the roadway, it must have been caused by the adjacent builders.

As a result, my attorney filed a claim against the sewer and water connection subcontractor – the only party commissioned by my *LLC* to open the street. Now there were four defendants. And the more parties involved, the more correspondence, the more legal maneuvering, and the more attorney fees. Meanwhile, the plaintiff was operating under a *contingency agreement*, which meant no legal fees for him. If any money was awarded, his lawyer would take a percentage, usually about a third. That structure gave the

283

284

285

286

287

288

plaintiff no incentive to give up, while the financial burden kept growing for the defendants.

I felt sorry for my sewer subcontractor. They had done a thorough job. When they backfilled the trench after completing the water and sewer connections, they compacted the soil as required. They were not allowed to finish the street with asphalt until a *city inspector* had approved the work. The *inspector* showed up, checked the site, and granted permission to proceed. Logically, once the city inspected and signed off, responsibility for the roadway should have shifted back to the city. So why was the city so quick to shift blame onto others?

It would make far more sense for the city to make a principled decision about its liability for maintaining public roadways. The countless potholes across New York show how widespread the problem is, and they must cause numerous accidents and lawsuits every year. Addressing liability as a matter of principle would be more efficient than endlessly pointing fingers.

What makes this case even more frustrating is the plaintiff's conduct. Common sense would suggest that his reckless behavior should disqualify his claim. Why was this lawsuit not dismissed immediately by the court? My conclusion is that the system is designed to serve the plaintiff and, even more so, the legal industry, which profits by keeping processes alive.

This lawsuit has now dragged on for several years. Every few months, the lawyers gather at court for what is called a *court conference.* I attended several of these, even though technically I was not supposed to (by then, my first lawyer had already stepped away). Most of the time was spent waiting. At the end of the session, a handwritten one-page document *(NYSCEF Doc. No. 48)* was issued, signed by the judge. It listed outstanding tasks and scheduled the next conference. Whether those tasks were ever actually addressed seemed irrelevant – the process simply continued.

At these sessions, there were usually about fifty lawyers in the room for a dozen cases. Each lawyer likely billed their client for two or three hours, even though most of that time was spent simply waiting. This inefficiency could easily be avoided if time slots were assigned, or if the meetings were conducted digitally, eliminating the need for travel altogether. At first, I did

289

290

291

not understand how these conferences worked until I began receiving bills from my legal representative. The charges for his time at these conferences added up to thousands of dollars, but they did not have any efficient outcome.

Another source of delay and uncertainty was the frequent change of presiding judges. I never understood exactly why this kept happening, but I believe it had to do with promotions or reassignments within the system. Each time a new judge was appointed, the case had to be explained again, and more time was lost.

It became increasingly clear that lawsuits like this are primarily about extracting money from *insurance coverage*. In New York State, the law allows a judge to assign partial responsibility for damages. The monetary award is then determined by the percentage of responsibility allocated to each party. This means that even if a judge considers a case to be partially *frivolous*, the plaintiff may still receive compensation. That means I am at risk of losing money, and on top of that, I must pay attorney's fees. It feels as though I have been forced into a situation where I carry all the disadvantages of being involved, without having any real control over decisions.

The clock keeps ticking. You can imagine how often I woke up or went to bed with this heavy *state-of-mind*, trapped in a cycle that seemed impossible to escape.

Is this fair? I always believed the legal system existed to serve humanity as a whole, with everyone treated equally before the law. I have lost not only large sums of money but also confidence and trust in a system that should protect all of us. Instead, it seems to be exploited by those who understand the process better than the people they draw into it for their own gain.

The architect, in a broader sense, is traditionally seen as someone skilled at shaping structures, policies, and systems, envisioning and guiding processes in ways that improve them. After many years of experience, I know that this idea is an exaggeration. But have we become so removed from realities outside our immediate focus that we ignore the obstacles we should recognize? If we consider ourselves *masters of process*, shouldn't we at least understand the barriers that stand in the way of achieving superior outcomes?

I am convinced that the way we understand ourselves as architects must expand toward greater comprehensiveness. I

292

293

294

295

296

297

298

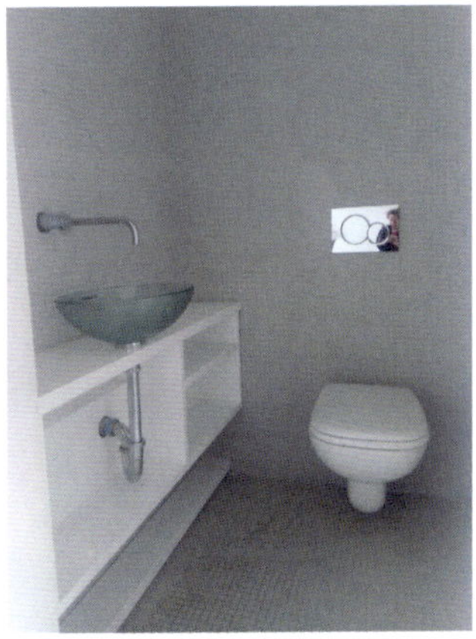

299

300

301

302

303

had never heard of such legal troubles before, but it changed the way I see the world. Could I have been more knowledgeable? In medicine, peer-reviewed journals disseminate information about unusual cases, professional experiences, and improved methods of communication.

In architecture, however, most publications lean toward self-promotion, celebrating beauty while avoiding any discussion of substantial problems.

If we want to be considered an *applied art*, we must pay attention to the forces that can negatively affect our performance. If publications addressing these issues exist, I am either unaware of them or they have been presented in such overly technical ways that I chose, regretfully now, not to engage with them.

As a teacher guiding architecture students into the profession, I also wonder how these concerns could be woven into the curriculum. The desire to improve the world aesthetically does not prepare you for the impact of certain laws. They may be called *justice*, but they are not always *just*. As the judge in my case put it, *"That's the cost of doing business in New York."*

304

305

306

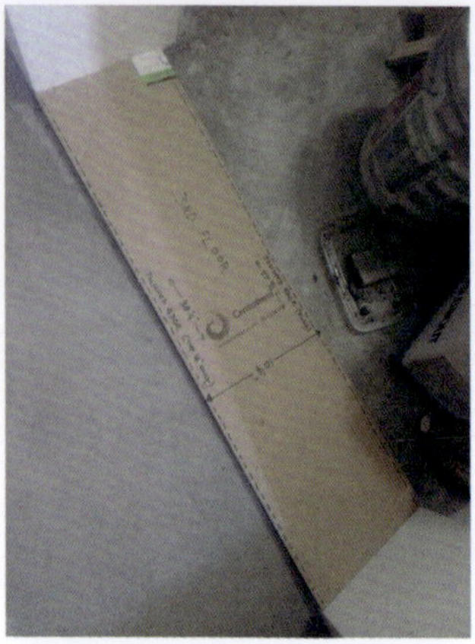

307

308

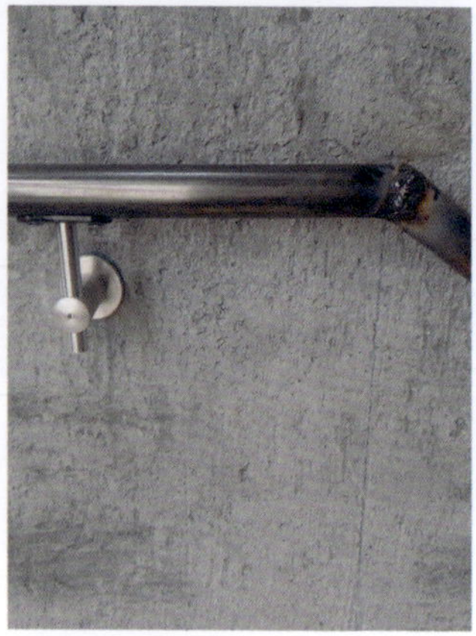

309

310

311

312

313

314

315

ACADEMIA

316

317

318

An architectural career usually begins in an academic setting, which should serve as the foundation for learning how to create good architecture. The discipline depends on *comprehensive thinking* that balances creative visions with economic constraints, technical and material knowledge, legal conditions, and the psychology and leadership skills necessary to connect all of these aspects. This raises an important question: does the academic curriculum truly prepare students with such a broad range of skills? Are we, in fact, prepared and equipped to teach *applied art* ?

In my more than twenty years as a professor of architecture, I have seen that most curricula do not cultivate applied skills. They are largely focused on the solitary acts of thinking and representing. *Conceptual thinking* often forms the center of architectural education, emphasizing the artistic idea rendered on paper, while the study of practical *means and methods* is marginalized. Yet combining these two is essential for developing an aspiring architect who wants to create tangible, built works. It is also critical for making a smooth transition from the academic environment to professional practice. If the goal were to achieve this balance, faculty would need to be selected for their ability to unite design thinking with practical expertise. This is the foundation of a *Design-Build* curriculum, which at our school is consistently named by students as one of the most popular topics. They mention that it also provides targeted improvement to their architectural education in ways that traditional studios cannot.

However, *Design-Build* projects require more time for teaching and procurement. They carry greater risks than projects that remain on paper, since so many things can go wrong during execution. As a result, many faculty members avoid such projects, often lacking the professional background needed to run them. I have even seen a countertrend, where professors focus narrowly on *research*, *theoretical work*, or *community activism*, with little or no tangible architectural output. Academic demands reinforce this tendency; a published book can count more toward career advancement than the realization of a built project. To me, it should be the opposite, since the tangible creation of an *applied art* lies at the heart of the discipline.

We must bring students closer to understanding architecture through hands-on experience, while also teaching them to see architecture as a *comprehensive* discipline that demands oversight

319

320

321

and knowledge across many different fields and specialties.

One of the disheartening comments I sometimes hear from alumni who hire recent graduates is, *" ... a good worker, but we had to reteach them everything..."* This is an indicator for a flawed curriculum. The issue lies in how academic training connects, or fails to connect, with the realities of professional practice. The overemphasis on *conceptual thinking* should give way to a more integrated approach, in which the abstract idea is tested *hands-on*, refined, and strengthened through the learning of concrete *means and methods* that serve the creative vision.

Often in academia there is an absence of *making*. Engaging in architectural full-scale fabrication is seldom practiced in an academic setting. Engaging the world of professional builders (who could absorb certain deficiencies in such experiments) is also commonly unlikely. However, it would change the way students approach architecture as a profession. The worlds of *thinking* and *making* must be connected, but this seems to be a difficult undertaking in a school environment. There were certain meaningful connections that I experienced while studying. For instance, at the Technical University Berlin it was a requirement for architecture students to document proof of several weeks of internships spent on construction sites, before being eligible to an academic *benchmark* called *Vordiplom* (aka.: *Pre-Diploma*). The reason for this was *"to give students an early insight into the fields of activity of the other people responsible for the implementation of a construction project and to give them an impression of the complexity and dependencies of the necessary work processes."*[18] As a result, I have worked with masons and I have worked in a construction management firm, both during summer breaks, early in my academic years. It was the first time I experienced different perspectives in construction, often perspectives that made me adapt my *thinking* on how architecture is brought into existence.

There are two essential elements in an architectural education when it comes to producing tangible architecture: *process* and *product*. The *product* is well covered in most architectural programs. We focus on architectural narratives, typologies, and the articulation of final appearances. We are analytical on paper, and we know how to represent *idea* precisely. However, we are far less effective

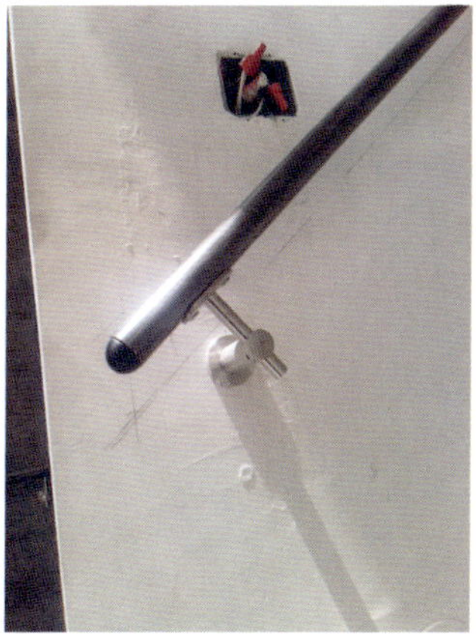

322

323

324

at teaching how to integrate the conditions of production. We, as teachers, often leave the students with the impression that *someone-else-will-figure-it-out*, which is seldom the case, and, if that someone can even be found, it will likely cost considerable money.

The way we teach *process* is either suppressed, or often theoretical, making it difficult to detect its crucial connection to the *product.* To truly understand *process*, one must enter the world of the builder and experience firsthand how things are made.

There are many opportunities within academia to broaden the curriculum and strengthen how we teach the significance of *process.* The most important step would be to establish a permanent *design-build* program within the curriculum, similar to other concentrations we have seen, for instance with the rise of Landscape Architecture. The best chance to develop such program is to invite the building world into the school's environment on a consistent and long-term basis. This means forming lasting partnerships with construction industry stakeholders and integrating these collaborators into the design curriculum.

Students are naturally drawn to the design component of their studies. It carries the highest number of credits (and therefore the most teaching hours). It is reinforced by the permanent cliché that an architect's most valuable skill lies in generating design ideas. This emphasis contributes to their affinity for design, but it also leaves a gap in their understanding of how those ideas are brought to life.

Opportunities exist, for instance within construction technology courses, but the hours attributed to such courses is often too low to make significant progress.

One can make students learn at construction site visits. While these can be difficult to organize due to liability concerns and construction scheduling, they offer valuable learning experiences. The problem is that site visits often position the architect as a bystander, which reinforces the mistaken belief that architects should remain removed from the construction *process.* In reality, the opposite is true: in order to evaluate the *product,* one must understand the *process.* For this reason, design-build, embedded in the *Design Studio* curriculum, is perhaps the most urgent area for development in today's architectural education. Rather than letting students believe

325

326

327

328

329

330

331

332

333

their work ends when drawings are complete, we should encourage them to think through a project's buildable anatomy from start to finish and actually engage them in building mock-ups for their understanding. In a world increasingly shaped by industrialized production, architecture must retain its connection to craftsmanship if it is to survive as an *applied art*. And where better to plant the seeds of this disposition than at the cradle of education, the academic setting! There has to be plenty of opportunity for future architects, so that they can thrive through their engagement with materiality.

Of course, not all students share the same interests, and the same is true for faculty. In my own department, I have noticed that few practitioners are hired into full-time positions. My suspicion is that the search committees, usually composed of current faculty members, will not feel at ease with bodies of knowledge they cannot significantly contribute to. Through this tendency the theoretical pathway, which is more difficult to be proven right or wrong, leads to less conflict among faculty. This reflects a broader withdrawal within academia from engaging directly with the *making* of architecture.Who should then be teaching *design-build* courses? This remains an open question. As I've mentioned before, these courses demand a significant time commitment and carry the high risk that the project will not be completed within a semester. They also involve greater supervision and responsibilities, particularly in managing safety and preventing injuries. They need greater funding, large tool sets, more coordination, and as a consequence greater leadership, with fundraising and project management skills. The higher effort must be solidly supported in a sound curricular structure, in order to promise success. The rewards can be extraordinary. Witnessing students grow as they gain hands-on knowledge related to their design skills is one of the most gratifying experiences in teaching. I have seen students transform dramatically, becoming more confident, more disciplined, and more deeply focused on understanding architecture from conception to completion. This is an almost immediate reward.

The larger payoff comes when these students enter their profession. Their ability to use *process* to develop a superior *product* will continue to expand over their careers, compounding for a lifetime and connecting to the practitioners' challenges. This is essential to our profession. If we aim to create *applied* art, we must engage in developing *artisanal* skills.

334

335

336

Throughout the five years of my own construction project, I spoke often to my students about what was happening on my construction site. I showed them progress photos in lectures. I offered tours on site, whenever they were interested, and most students showed real curiosity, asking thoughtful questions not only during the visits but even months afterwards. A few became so interested that they returned to help me with small, hands-on tasks. Two of them even obtained their NYC-certified 40-hour safety training that made them eligible to regularly work on site, for instance during summer breaks.

During site visits, I always tried to facilitate conversations between students and professionals, who were present during that time so the students could better understand the nature of on-site discussions and the range of viewpoints. It also became clear, even in later semesters, that for those who were curious, it was easier to grasp what it truly means to draw. For example, they came to understand that gravity is not just an abstract concept on paper. Once, we carried a heavy water heater up to the roof, it was immediately apparent that logistical challenges exist that seem deceptively simple in a drawing.

Some students realized that the two lines representing a pane of glass in a drawing could translate to more than 150 pounds of actual weight. Others learned, on the job, about the need to prevent electrochemical reaction between dissimilar metals in a facade. Some discovered the correct method for gasketing a curtainwall system, using special tools to trim the overlaps between horizontal and vertical rubber elements according to specifications. Others learned about varying consistencies in concrete mixes – and about the sheer effort required to carry those heavy bags up multiple flights of stairs.

I also shared drawings with them. But these were not the polished, rendered *visions* that many students might have expected. In academia, much time is devoted to digital representation, especially to visualizing spatial environments. The quality of such work has advanced tremendously; today, it can be difficult to distinguish between a rendered image and an actual photograph. However, what has suffered in architectural education are the organizational skills needed to document assemblies precisely, with coherent and accurate dimensions. For many students, the visually exciting world of rendered images holds more appeal, at least at first, while the understanding that precise documentation is the primary tool for achieving the desired result develops often much later.

My own drawings were simple black-and-white technical documents, focused on carefully balancing the *vision* with the specific materials and assemblies required. Everything was pragmatically dimensioned, with material take-off information, so that the fabricator would know exactly what materials to order, and thus relate it to costs. These were straightforward line-drawings, most often orthographic projections in plan, sections, and elevations, supplemented with close-up details referenced back to the base drawings. Choosing the right architectural scale was essential to communicate both the design and the assembly efficiently.

337

This is the point that is often lost on students: these *unappealing* line drawings are not simply a means of showing how assemblies are built. They are an essential design tool, a way to test whether the concept is efficient, practical, and materially sound. They also serve as the clearest instrument for controlling the construction process. The more competently the drawings convey assembly strategies, the more likely the finished product will match the original vision. The drawing is testing the *process.* In this way, the true value of a drawing lies not in how beautifully it imagines what something might look, but in how clearly it directs the making of something you already know will look and perform exactly as intended. This book contains many photographs documenting the *process* of this particular project. It does not include most of the drawings, because I believe the ability to solve problems develops most effectively and primarily through observing real things as they come into being. This kind of thinking grows from discoveries made in a world that already exists, and it stimulates the mind of the tinkerer. Drawings will follow. They are a simulation of that tinkering on paper; they test strategies rather than embody them.

338

In architecture, solutions are never identical, because each one is specific to its project. The magic we sometimes encounter in an architectural experience is tied to the tangible inventions we can articulate and bring into existence. This is a clear reminder that we must care about the *means and methods* of construction, even if certain advising institutions warn us against becoming too involved. In truth, there is no avoiding it: the best design results emerge when knowledge of *process* and *product* act together with the combination of *means and methods* and *appearance*. This is what I often communicate to my students, preferably on construction sites, where discoveries are concrete and tangible. In the *means and methods*, you often find the seeds of invention, the very elements that can unlock the magic of a fully realized architectural experience.

339

340

341

342

343

344

345

346

347

348

CONCLUSIONS

349

350

351

My personal conclusion first: I now know that I am capable of building a substantial house. I also believe that the building is reflecting its contemporary architectural principles and its related craft. One might even be able to sense a bit of the atmosphere and the extraordinary way it came into existence. For me that much is clear from this personal experience, which took me far longer than I ever imagined. To prove myself at a tangible level was my long-term goal from the beginning of my architectural studies, and I am proud I achieved it.

But more importantly, beyond the personal experience, the larger question is what this means for our profession as a whole. *How should architects conduct their work?* I am often asked whether I would approach the project the same way if I could turn back time. My answer is wholeheartedly: yes. But there really was no other choice. If I had attempted to run this project in the typical, conventional way, subdividing responsibilities into many different camps, I would have spent far more money than I had, or could borrow. Costs would have ballooned through *supervision fees*, *change orders*, and *profit margins* that increase as more parties are involved. The conventional system is set up for people with greater financial resources. By following that path, I also would have lost significant control over the *process* of actually *making* architecture.

What proved crucial, was learning to navigate the complexities of construction while constantly asking what additional roles I, as the architect, could assume in order to keep the budget lean. This is not the typical response of an architect. More often, they say: *not-my-job,* and step aside within the highly compartmentalized system of construction. I believe we must do the opposite. We should continually question what our responsibilities could be, and be prepared to take them on when it leads to a better result. Architecture can be created more meaningfully, with better methods, greater economy, and more sophisticated outcomes, if we are willing to take risks. Unfortunately, in recent decades architects have largely withdrawn from this. Our primary tool, the set of drawings, remains isolated within the *white-collar* world of *thinking*. The riskier *blue-collar* world of *making* stands apart, seeking partners who know how to run projects quickly and profitably. As a result, architects have become subsidiaries to developers, who now hold the comprehensive understanding of *process* that architects once claimed.

352

353

354

Architects, even though they hold broad and comprehensive knowledge, often excuse themselves from the everyday demands of the construction site. Many are reluctant to take charge of anything that carries significant risk. As a result, architects are often left out of the actual *process* of building. I even know developers who provide all architectural services themselves, often by reusing design elements again and again. Whatever documents must be *signed and sealed* for the Department of Buildings are handled by the structural engineer they retain. In such cases, there is no need to pay an architect at all.

The question is: why do we, as architects, restrict ourselves so much? Why do we not compete more directly by exposing ourselves to the same comprehensive strategies that developers apply? Through our education, we understand craft and fabrication methods, performance criteria, and we likely have a deeper sense of aesthetic value. I believe we could produce far better buildings if we learned to take greater control over a *process* that is, after all, named after our profession: *architecture*. The developer has become the mainstream figure, delivering predictable products at predictable prices. Developers also hold stronger control over certain essential aspects, such as securing financing, obtaining permits, filing paperwork efficiently, and negotiating with contractors and trades. Yet these skills are learnable, particularly for architects who are trained in *critical thinking*. What is needed is the willingness to understand the *rules of the game* and the curiosity to engage with them. Instead, for quite some time the architectural community has retreated into the office, observing construction mainly from a distance, until the project is complete.

I believe that our architectural profession is in a serious crisis. If we do not change the way we produce, we will make ourselves obsolete. Others will continue to take over our work and do it more efficiently. Our margins will shrink further, as new *players* assume the very tasks we once considered our strengths. To remain relevant, we must learn to understand the entire operation of producing architecture in a comprehensive way. We need to develop expertise in areas we have traditionally avoided, so that we can once again become the driving force behind the production of architecture, delivering outcomes that are not only aesthetically strong, but also economically, logistically, legally, and culturally sound.

355

356

357

Where do we learn this? The right strategies for surviving in a competitive market should be addressed in architectural education. Yet one of the problems is that the same competitive rules do not apply to teachers within the academic system. The priorities in academia are often not about preparing students for real-world conditions but about meeting the expectations of leaders within the system. Instead of being rewarded for building beautifully, faculty are rewarded through publications, conferences, and recognition in national and international academic circles. These cannot be the right tools to respond to a profound professional crisis. But if academic isolation is driving the system, how much influence can it still have on the profession?

For me, the most important task is preparing the next generation of students to question the current setup. In my role as professor, I am less concerned with teaching specific topics and more interested in showing students how to apply their learning to real-world conditions. This means being observant, street-smart, flexible, and able to adjust strategies dynamically to particular environments, while still producing work that inspires. The entrepreneurial qualities of the architect must include discovering synergies, building innovative models of collaboration, and carefully monitoring conditions as they evolve. The results of this approach are likely to be far more interesting than the standardized *Model-T-strategy* of the developer. Above all, it is about developing confidence: understanding the conditions, but also thinking independently, with a mindset that looks for leverage in the real world of *making*.

The emphasis must be on becoming the master of *process* – understanding, better than anyone, the specific conditions that apply and using that knowledge to shape the artifact, with an improved placement strategy for the final *product*. We have the potential to develop improved methods of the process and the ability to exert greater control over architectural outcomes. What is required is learning and teaching how to apply these methods in a meaningful way. The curriculum must evolve to introduce activities that are practical and focused on preparing architects to become true masters of construction in service of creating beautiful and inspiring design. A superior product depends on a deep knowledge of the conditions under which it is produced.

358

359

360

361

362

363

364

365

366

367

368

369

370

371

372

373

374

375

376

377

378

Some of the institutionalized models we follow today – professionally, such as the AIA *A201* contract template discussed earlier – remain static in their generic approach to construction. They have been in this state for a long time. These contract templates are reviewed every ten years, most recently in 2017. Yet the same passage appears as it did thirty years earlier, in 1987: "*the architect will not have control over or charge of and will not be responsible for construction means, methods, techniques, sequences or procedures.*"[19] The world has changed dramatically since then, but the language has not. Even though there are alternative templates, such as the AIA *A141* for *Design-Build* projects, the prevailing stance remains essentially one-model-fits-all. The dominant tone in many of these documents is about minimizing risk. Yet, as argued before, risk-taking is intrinsic to construction, and thus to architecture. Problems must be solved, beautifully, but also on time and on budget. Whoever does not contribute actively to managing a project during construction will sooner or later become obsolete. Generalizations are rarely useful when it comes to *making* architecture. Individual *strategic thinking* is required to take advantage of specific conditions. Architects must not only understand the risks that come with construction but also learn how to anticipate and engage them. Rather than prescribing rigid, inflexible rules that seem designed only to prepare for lawsuits, it is far more important to establish conditions that are helpful and productive for the entire team.

The architect of the present must be a leader and a team worker, *"a mason, who has learned Latin"*.[20] Any legal contractual *fire wall* will primarily not lead to avoidance of risk, but to obsoletion. However, in terms of education, exactly the conditions that might create risks must be taught and understood, so that architects, especially the next generation, can evaluate them comfortably and competently.

The main title of the book: "undercover architect" points to my conviction that certain necessary skills for *making* architecture can only be understood practically. It is much more likely that you uncover them in real life, involved in construction.

My role model for relearning such skills as an architect is the figure of *the Masterbuilder*,

379

380

381

described in the introduction, which I hope can eventually reestablish a tangible connection to the world of *building*. This might seem a bit nostalgic, but I believe that we have to have a holistic understanding of the building processes, from concept to completion, and oversee all aspects, including material selection, techniques, details, structural and mechanical integrity, artistic elements and spatial assets, financing, code compliance ... and much more. The role of the specialist for the architect has to be reconsidered, because the separation of professions led to our isolation. The integrated *All-rounder* holds in an economic sense greater promise than the *No-risk-bystander*, as suggested by the AIA's contract templates. Many tasks in construction can only be understood through material engagement, so one must seek or create such opportunities – whether through design-build projects in school, internships with construction firms, or practicing one's own makeshift building. With this hands-on approach, it becomes clear that our profession is truly a hybrid: a *blue-and-white-collar* discipline, with one foot firmly in each world. The common cliché is that architects never get their hands dirty, always seated comfortably in the driver's seat. I doubt that such a position is possible, or even desirable. You are missing out on all the fun: The necessary skills for *making* architecture can only be understood practically. It is much more likely that you uncover them in in a construction environment, which I therefore *infiltrated* for the *Townhouse in Harlem* project.

Reforming our approach means preparing the next generation of architects to understand their conditions differently. We must communicate more comprehensively, highlighting the crucial roles in construction and management that architects could potentially fulfill. The goal is not only to foster efficiency but also to cultivate synergy and empathy among all parties involved in construction. This includes compassion for the architectural artifact itself, shared from every perspective. To achieve this, we must create more opportunities for exchange with all construction stakeholders by forming a stronger, more collaborative community. Many processes could and should be practiced together. The business and management expertise of the developer, the craft and installation techniques of the builder, the knowledge of code experts, the understanding of legal frameworks specific to

New York, and the design vision of the architect – all these elements must be integrated into a more comprehensive education.

Ideally, this would mean that developers, general contractors, skilled builders, and architects could study under one roof, teaching one another and exchanging knowledge. In some ways, this already happens unknowingly. Many of the students I have taught have become construction managers, others have become examiners at the NYC Building Department, some have become fabricators specializing in sophisticated processes and finishes, while others have pursued the path of more traditional architecture. Yet, most of them did not realize during their studies that such different paths existed, or that they might one day be drawn to a particular way of thinking. Much of this happens through coincidence, often outside academia. Sometimes it grows from students' self-initiated extracurricular activity and their hunger for new experiences. It also means that there are networks of expertise: you should know how to connect different experiences and make them accessible: share your expertise and tap into networks!

For those who aspire to become *design-builders*, who want to connect their design thinking directly to fabrication, academia has not yet advanced enough to provide a systematic curriculum, at least not to my knowledge. There are brilliant historic precedents, such as the Bauhaus, with its clear vision of unifying craft, art, and technology, and of integrating industrial principles to produce extraordinary artifacts. Frank Lloyd Wright's Taliesin West and Paolo Soleri's Arcosanti might be good historic examples of *Design-Build* fellowships, where you could experience a sophisticated architectural idea from the architects drawing board to its *means and methods* of realization. A few very good design-build programs stand out today, among them: Auburn University's Rural Studio program, University of Kansas Designbuild program, the Yale Architecture Building Project, or Brian MacKay-Lyons' independently organized Ghost Laboratory.

So far, much progress in this area has come from students' own initiative and their instinctive push toward hands-on knowledge. Natural curiosity about tangible things is a prerequisite for being interested in architecture.

382

383

384

That curiosity, however, should not be left only to chance. It must be embedded in education. We should teach architecture with far greater focus on application and on cultivating ways of thinking that match the *processes* of *making*. The true proof of our value as architects lies in producing results that are demonstrably better than what speculative development alone provides. Only then can we build trust in the profession and affirm the cultural contribution it is uniquely capable of making.

Often unexpectedly, students thrive after school in environments that suit their talents, combining the creative skills of an architect with the entrepreneurial qualities of a businessperson. Their way of thinking changes as they engage with complex processes, bridging the worlds of design and construction. If this is true, then our educational system must adapt. We need to introduce topics into the curriculum that foster entrepreneurial qualities, while also ensuring that tangible skills in craft, art, and technology are practiced *hands-on*. This is especially true for architecture, though one could ask the same of any field: If you want to become a surgeon, shouldn't you begin learning certain practical *hands-on* skills in college? Perhaps not yet on a human body, but at least by practicing on a frog – or in New York City, maybe on a rat.

In many European countries, the cabinetmaker's apprentice must, in the final year of a three-year apprenticeship, build a *Gesellenstück*, a complex, self-designed artifact that demonstrates advanced proficiency. Architecture is more complicated. Most of what is taught remains abstract and too often unapplied in an academic setting. But imagine that at the end of your architecture studies you would have to build a full-scale mock-up of a corner of a house you designed.

Drawings, even though they are academically emphasized, seldom connect directly to tangible construction. The closest link might be model building, but even there, models are often too small and often painted white, deliberately stripping away their materiality, when they could instead become opportunities for engaging with construction and craft.

Technology courses sometimes take this further by building larger models that simulate

385

386

387

actual construction. Mock-ups of key details are also occasionally attempted, linking material directly to overall design intent. These projects are sometimes criticized by students or faculty as being *over-the-top*. In my view, they are essential. They offer an understanding of *materiality* that is fundamental to architecture.

Building something with construction in mind rarely happens in academic contexts. Instead, students produce countless drawings, seldom technical and often illustrative, that signal a lack of commitment to true materialization. Representation is too often treated as a form of entertainment, with drawings functioning as follies that serve aesthetic pleasure rather than practical development.

How, then, can we transition to a profession that depends on *applied thinking*? One approach is to create new opportunities within the curriculum for practicing *means and methods*. Another is to develop parallel avenues outside of school, such as internships that focus on tangible skills. With such practical credentials, the built environment, shaped by architectural thinking, would likely flourish. Young architects would gain confidence in their understanding of how things are put together and become more comfortable engaging with the world of *materiality*.

I admit I am torn on the right strategy. Of course: Ideally, this should be integrated into a strong curriculum that reaches many students. But is this realistic, given the resistance and problem-solving such change requires? I am not sure. Reflecting on my own development as an aspiring architect, I have to admit that much of my education lacked a suitable environment for engaging with *materiality*. Most of what I learned in that area came from my own initiative and curiosity. Yet what makes a movement strong is the initiative of many. My hope is that the next generation of architects, including the students I have had the pleasure of teaching, will approach this curiosity with openness and be encouraged by this book.

Architecture is a beautiful profession because education never stops. There is always more to learn, and it never becomes boring. For me, construction site visits remain a favorite. They are consistently eye-opening, offering new strategies to observe and new ways of

388

389

390

understanding how those strategies can shape architectural space. (In that respect: this book is actually kind of a construction site visit.) Yet while there is so much to learn, there is too little access to applied knowledge of *means and methods*. The architectural community often appears unsupportive and too critical of one another. That must change. We need a culture that is more compassionate and collaborative, one that shares real experiences and builds broader community. Imagine, for example, if architects' names and addresses were placed on *Work in Progress* signs. This simple gesture could foster greater interdisciplinary exchange, encourage participation, and make visible the knowledge that shapes our built environment.

I also realized that architects, if really passionate about the aesthetic outcome of a design cannot think like the other *players* in this organizational structure. We cannot just be pragmatic and logical, we sometimes have to take on a more uncomfortable and extraordinary position: Not seeing the world for what it is and instead preferring to always imagine that it could be something different, something superior. That suggests that you cannot just follow the average *means and methods* that are proposed to you. You also have to be inventive and curious about discovering feasible alternatives that are able to enhance the artifact. Developing them is tough, because at the same time you must be concerned about the effort to construct them: Comprehensive practice brings you closer to perfection.

This project, the *Townhouse in Harlem*, was unusual in the way it was organized during construction. The more time I spent with it, and the more I invited students, colleagues, and friends to visit and discuss its organization, the more I felt that we, as architects, are underselling ourselves. We could take on a far greater role in construction than is commonly assumed. If I had approached the building's construction in the conventional way, the following positions would have been necessary: *construction manager, construction superintendent and administrator, project manager, site security,* and multiple *part-time assistants* for various trades, amounting to hundreds of hours of labor. Then there is the daily clean-up person and *general laborer*, who takes care of moving things around, removing garbage, and making sure that the jobsite is safe. In fact, that person often has some of the

391

392

393

394

395

396

best oversight of a project, since whatever is left unfinished at the end of the day usually falls to them unless clear and immediate instructions are given to prevent extra work.

All of these *functions* I chose to take on myself, serving as the central point of contact on site and wearing *multiple hats*. I enjoyed this role because it gave me the ability to measure progress directly at the end of each day. It also allowed me to troubleshoot effectively. When things did not go as expected and change orders might otherwise have been required to resolve procurement issues, I was able to find solutions quickly without automatically resorting to additional costs. Discussing solutions directly with tradespeople not only saved money but also built knowledge and trust and led to more control over the project.

Had I turned the project over to a typical General Contractor, the budget would have increased by an estimated 150 to 200 percent, which I could not have afforded. Beyond that, countless additional hours would have been consumed by communication with many parties. Many avoidable meetings and questions would have been generated, adding to both direct costs and my own time commitments.

This is a clear indicator: The proficient Architect must keep a clear head observing and contributing to all the necessary functions of a project, and must be willing to wear and change as many of those *hats*, based on the project's holistic requirements. The more you make them your own, the more proficient you will be in the world of architecture. The future of good architecture lies in *design-build* management held by a single hand. It should be the hand of the architect – and that is an architect with an expanded repertoire and set of responsibilities. This strategy is more efficient, it strengthens communication, and it leads to better results. It also results in greater control in aligning the original design intention with *real-world* possibilities. Yet, it requires both attention and diplomacy to navigate the system without stepping on the toes of others.

When I came to this country, I quickly realized that different environments favor different *means and methods*. Building techniques and materials that were common in Germany were not readily available in the United States.

397

398

399

For example, in Germany, many buildings are constructed in concrete or masonry, with far less reliance on steel framing. Here, the situation is often reversed. Many single-family houses in Germany are built from a particular type of masonry block that does neither exist nor does the name even translate well into English. It takes time to find the right *recipe* within a certain culture. It took me a while to understand the *ingredients* that were available and that I wanted to use in order to make something beautiful. A developer would likely not think this way. They so often rely on what has always worked, avoiding experimentation, and trusting mostly what has a proven track record.

But using the same *ingredients* always leads to the same outcome. Day after day, repetition sets in, and people eventually tire of it. Is this also true in New York? It seems that the tolerance for what gets built here is low. People appear resigned to the idea that architecture primarily serves the purpose of making money, with aesthetics coming a distant second. Where does this indifference come from? Perhaps it stems from a lack of basic education, maybe even from a lack of excitement, from a cultural environment that has lowered its expectations, and from the separation of stakeholders who rarely pursue comprehensive objectives together.

From a technical standpoint, the New York City environment is both unique and highly challenging. You must adapt your thinking to many different influences and impacts. You must analyze each situation from multiple viewpoints to understand how the various parties and the services they provide fit together. The past chapters document my thinking in this one particular case. The players in the world of architecture are familiar, but the conditions of interaction are never the same. Interaction itself is the key to making progress in the right direction. The opportunity for architects today is to reinvent the profession by becoming true leaders of interaction across all the different realms that shape a project, especially on the construction site. We should awaken from the slumber of the office and re-engage with the world that produces the built work. That field is where the architect's presence has been lacking for decades. One of the most common abbreviations on an architect's drawing reveals this attitude. *V.I.F.* stands for *verify in field*. It pretty much expresses the *not-*

400

401

402

my-job attitude again. But if we are to be in charge of the *process*, it would be far better if we felt the responsibility of verifying conditions ourselves, *in field.*

We must improve in order to interact, communicate, and contribute more effectively. During interviews with contractors of all types, it is essential to be thorough, always trying to sense how communication will work between parties. The contractor's main objective is usually efficient time management. The faster and more expediently a job can be completed, the lower the risk. For us as architects, perfection of our *vision* may be one of the main key objectives. To meet at a productive point in-between, we must understand the contractor's concerns and compare them with our architectural intentions. This is only possible if we are competent in seeing and respecting other perspectives. It is difficult to estimate real costs and the quality of operations before a project begins. By taking on more related responsibilities, we can build true competency in this field. This is where risk-taking becomes essential. Either we accept risk and learn how to strengthen our skills on the tangible side of the profession, or we remain fixed on the abstract side, where eventually we risk becoming obsolete.

Everybody loves good music. Everybody enjoys good food. Both are closely related to architecture in their creative and compositional nature and in our appreciation of beauty. They bring comfort because we are experiential beings. We delight in the thoughtful sensation of sound, the culinary pleasure of taste, and the visual symphonies of light, geometry, material, and space that shape our everyday lives. It influences our thoughts, our feelings and our culture.

It is worth every effort to preserve and reform a profession that contributes so much to beauty and human experience. If architecture becomes nothing more than an alibi for industrial mass-production and economic value, its true power is diminished and its beauty muted. What is needed is for us, as architects, to develop the skills required to create formidable, tangible artifacts that are at once stimulating and beautiful, for everyone.

Design-build may be the key to rediscovering these quality assets.

403

404

405

406

407

408

409

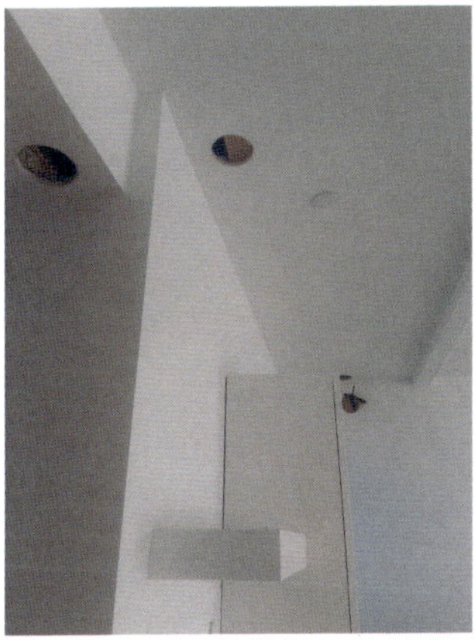

410

411

412

413

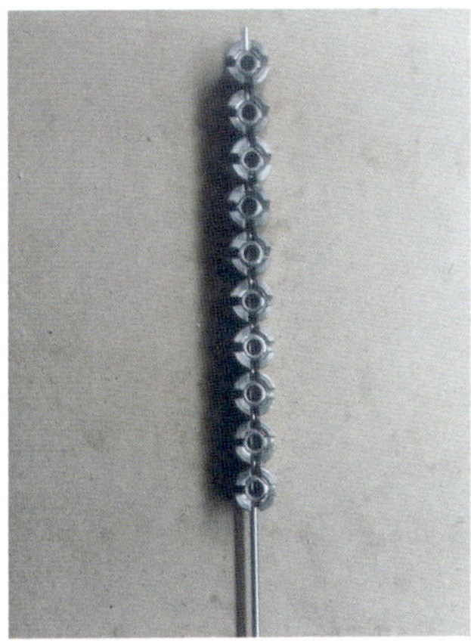

414

415

416

417

418

419

420

421

422

423

424

425

426

COMPLETION + PROCESS

** Photos by Albert Vecerka, ESTO

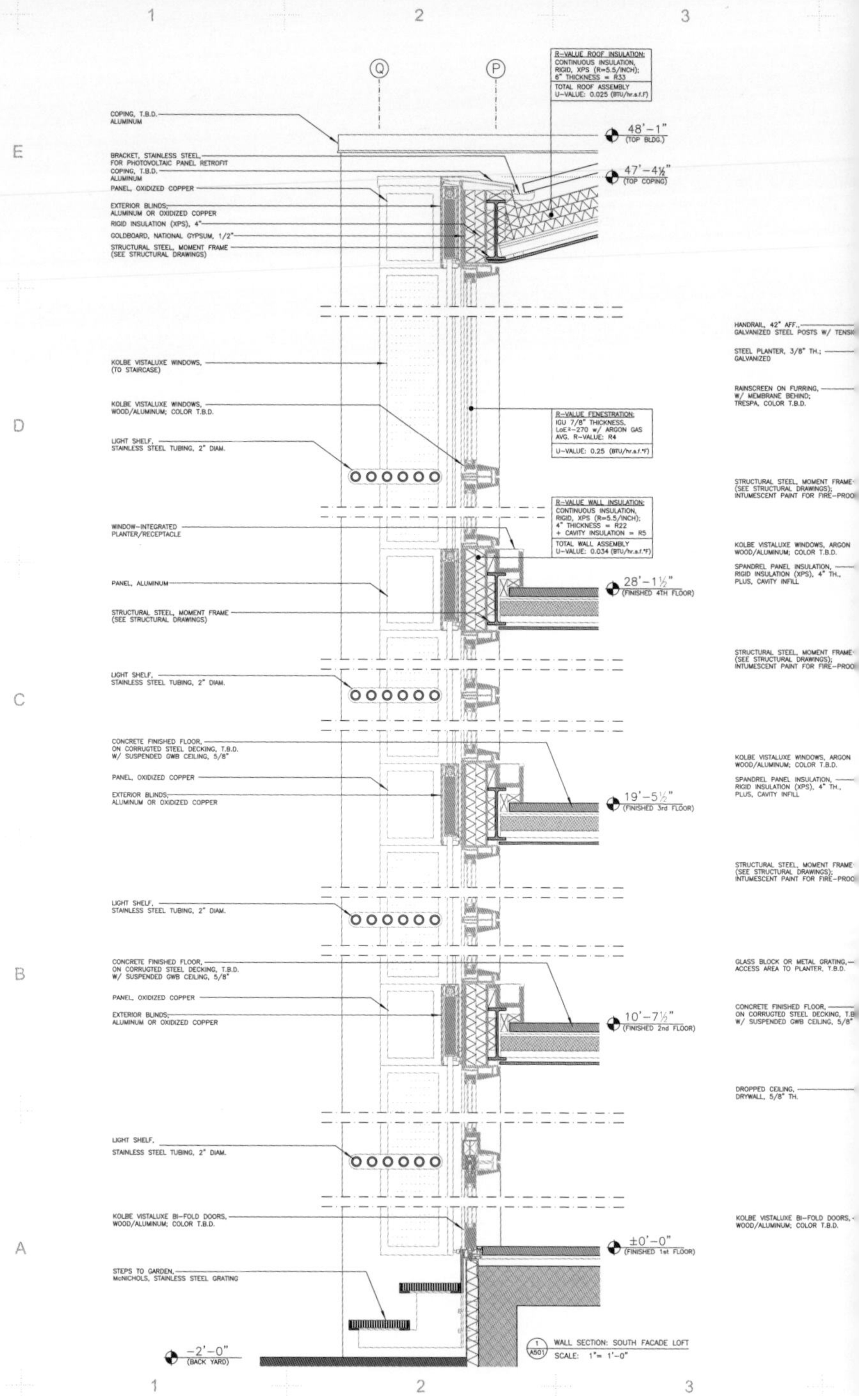
1
2
3
Q
P
E
D
C
B
A
COPING, T.B.D.
ALUMINUM
BRACKET, STAINLESS STEEL,
FOR PHOTOVOLTAIC PANEL RETROFIT
COPING, T.B.D.
ALUMINUM
PANEL, OXIDIZED COPPER
EXTERIOR BLINDS;
ALUMINUM OR OXIDIZED COPPER
RIGID INSULATION (XPS), 4"
GOLDBOARD, NATIONAL GYPSUM, 1/2"
STRUCTURAL STEEL, MOMENT FRAME
(SEE STRUCTURAL DRAWINGS)
R-VALUE ROOF INSULATION:
CONTINUOUS INSULATION,
RIGID, XPS (R=5.5/INCH);
6" THICKNESS = R33
TOTAL ROOF ASSEMBLY
U-VALUE: 0.025 (BTU/hr.s.f.F)
48'-1"
(TOP BLDG.)
47'-4½"
(TOP COPING)
KOLBE VISTALUXE WINDOWS,
(TO STAIRCASE)
KOLBE VISTALUXE WINDOWS,
WOOD/ALUMINUM; COLOR T.B.D.
LIGHT SHELF,
STAINLESS STEEL TUBING, 2" DIAM.
R-VALUE FENESTRATION:
IGU 7/8" THICKNESS,
LoE²-270 w/ ARGON GAS
AVG. R-VALUE: R4
U-VALUE: 0.25 (BTU/hr.s.f.°F)
R-VALUE WALL INSULATION:
CONTINUOUS INSULATION,
RIGID, XPS (R=5.5/INCH);
4" THICKNESS = R22
+ CAVITY INSULATION = R5
TOTAL WALL ASSEMBLY
U-VALUE: 0.034 (BTU/hr.s.f.°F)
WINDOW-INTEGRATED
PLANTER/RECEPTACLE
PANEL, ALUMINUM
STRUCTURAL STEEL, MOMENT FRAME
(SEE STRUCTURAL DRAWINGS)
28'-1½"
(FINISHED 4TH FLOOR)
LIGHT SHELF,
STAINLESS STEEL TUBING, 2" DIAM.
CONCRETE FINISHED FLOOR,
ON CORRUGTED STEEL DECKING, T.B.D.
W/ SUSPENDED GWB CEILING, 5/8"
PANEL, OXIDIZED COPPER
EXTERIOR BLINDS;
ALUMINUM OR OXIDIZED COPPER
19'-5½"
(FINISHED 3rd FLOOR)
LIGHT SHELF,
STAINLESS STEEL TUBING, 2" DIAM.
CONCRETE FINISHED FLOOR,
ON CORRUGTED STEEL DECKING, T.B.D.
W/ SUSPENDED GWB CEILING, 5/8"
PANEL, OXIDIZED COPPER
EXTERIOR BLINDS;
ALUMINUM OR OXIDIZED COPPER
10'-7½"
(FINISHED 2nd FLOOR)
LIGHT SHELF,
STAINLESS STEEL TUBING, 2" DIAM.
KOLBE VISTALUXE BI-FOLD DOORS,
WOOD/ALUMINUM; COLOR T.B.D.
±0'-0"
(FINISHED 1st FLOOR)
STEPS TO GARDEN,
McNICHOLS, STAINLESS STEEL GRATING
-2'-0"
(BACK YARD)
1
A501
WALL SECTION: SOUTH FACADE LOFT
SCALE: 1"= 1'-0"
HANDRAIL, 42" AFF.,
GALVANIZED STEEL POSTS W/
STEEL PLANTER, 3/8" TH.;
GALVANIZED
RAINSCREEN ON FURRING,
W/ MEMBRANE BEHIND;
TRESPA, COLOR T.B.D.
STRUCTURAL STEEL, MOMENT
(SEE STRUCTURAL DRAWINGS);
INTUMESCENT PAINT FOR
KOLBE VISTALUXE WINDOWS, ARGON
WOOD/ALUMINUM; COLOR T.B.D.
SPANDREL PANEL INSULATION,
RIGID INSULATION (XPS), 4" TH.,
PLUS, CAVITY INFILL
STRUCTURAL STEEL, MOMENT
(SEE STRUCTURAL DRAWINGS);
INTUMESCENT PAINT FOR
KOLBE VISTALUXE WINDOWS, ARGON
WOOD/ALUMINUM; COLOR T.B.D.
SPANDREL PANEL INSULATION,
RIGID INSULATION (XPS), 4" TH.,
PLUS, CAVITY INFILL
STRUCTURAL STEEL, MOMENT
(SEE STRUCTURAL DRAWINGS);
INTUMESCENT PAINT FOR
GLASS BLOCK OR METAL GRATING,
ACCESS AREA TO PLANTER, T.B.D.
CONCRETE FINISHED FLOOR,
ON CORRUGTED STEEL DECKING,
W/ SUSPENDED GWB CEILING,
DROPPED CEILING,
DRYWALL, 5/8" TH.
KOLBE VISTALUXE BI-FOLD DOORS,
WOOD/ALUMINUM; COLOR T.B.D.
3
2
1

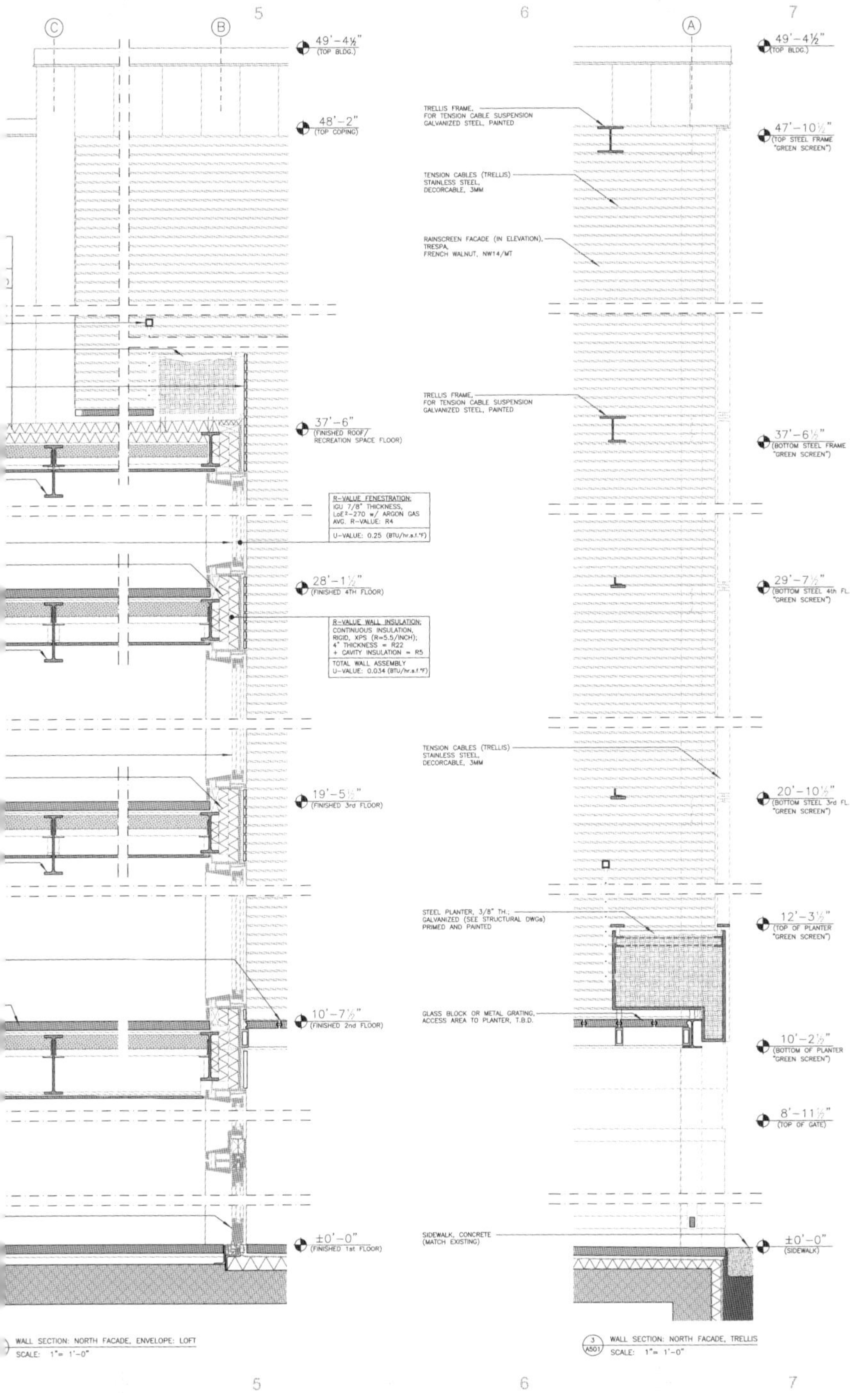

Main Façade: Wall sections + detail studies

106
108

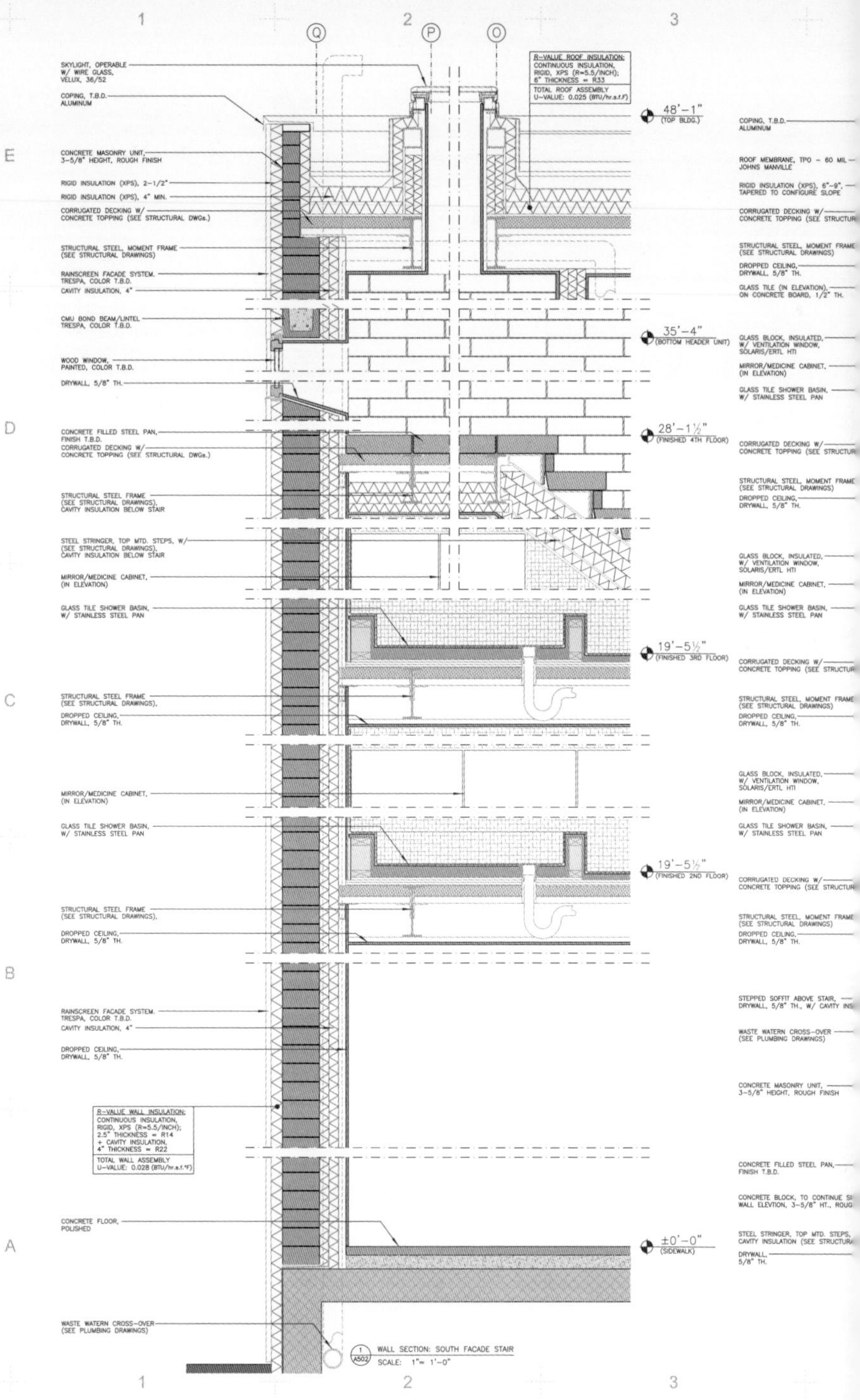

1 WALL SECTION: SOUTH FACADE STAIR
SCALE: 1"= 1'-0"

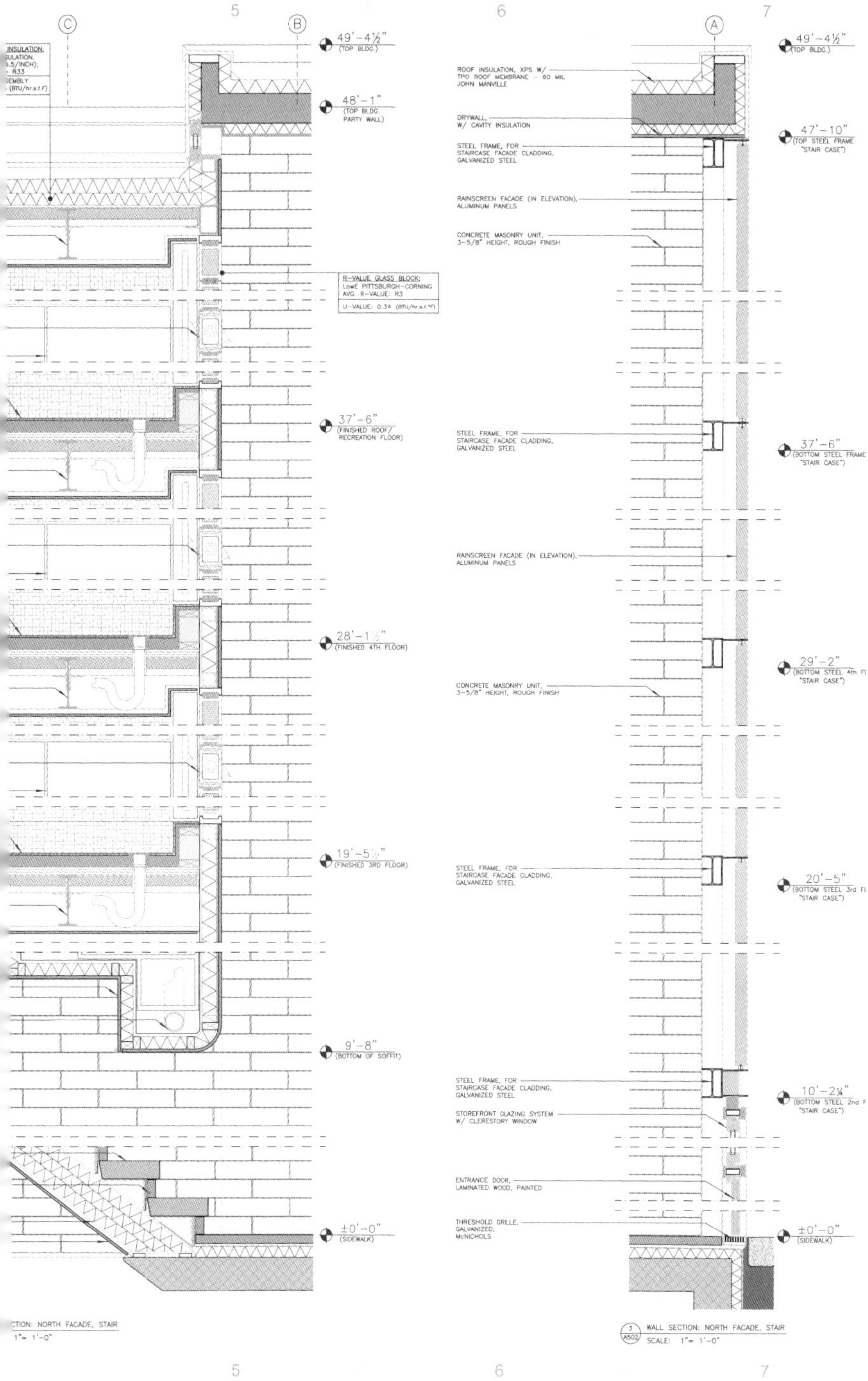

Stair case +
Stair Façade:
Wall sections +
detail studies

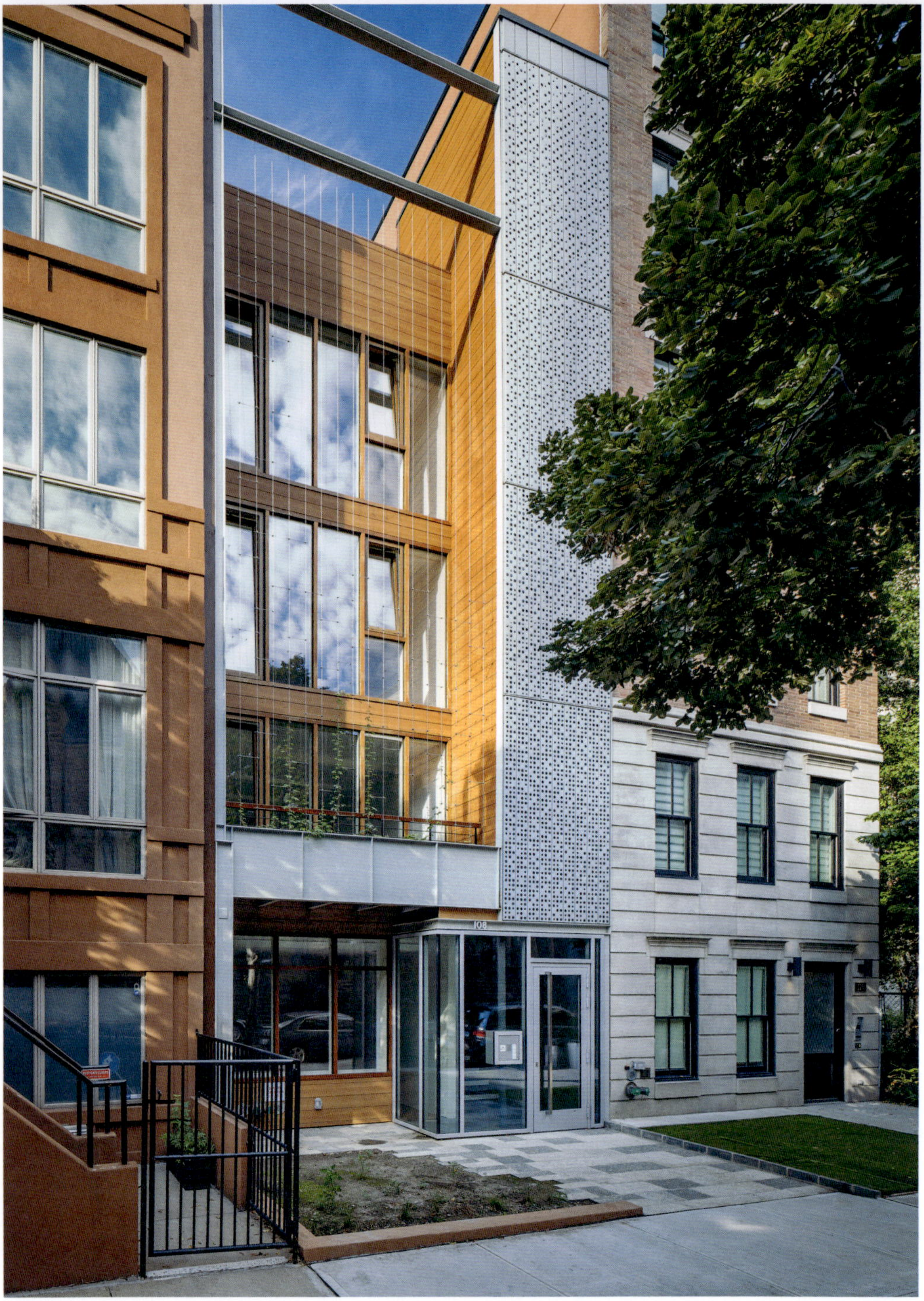
108

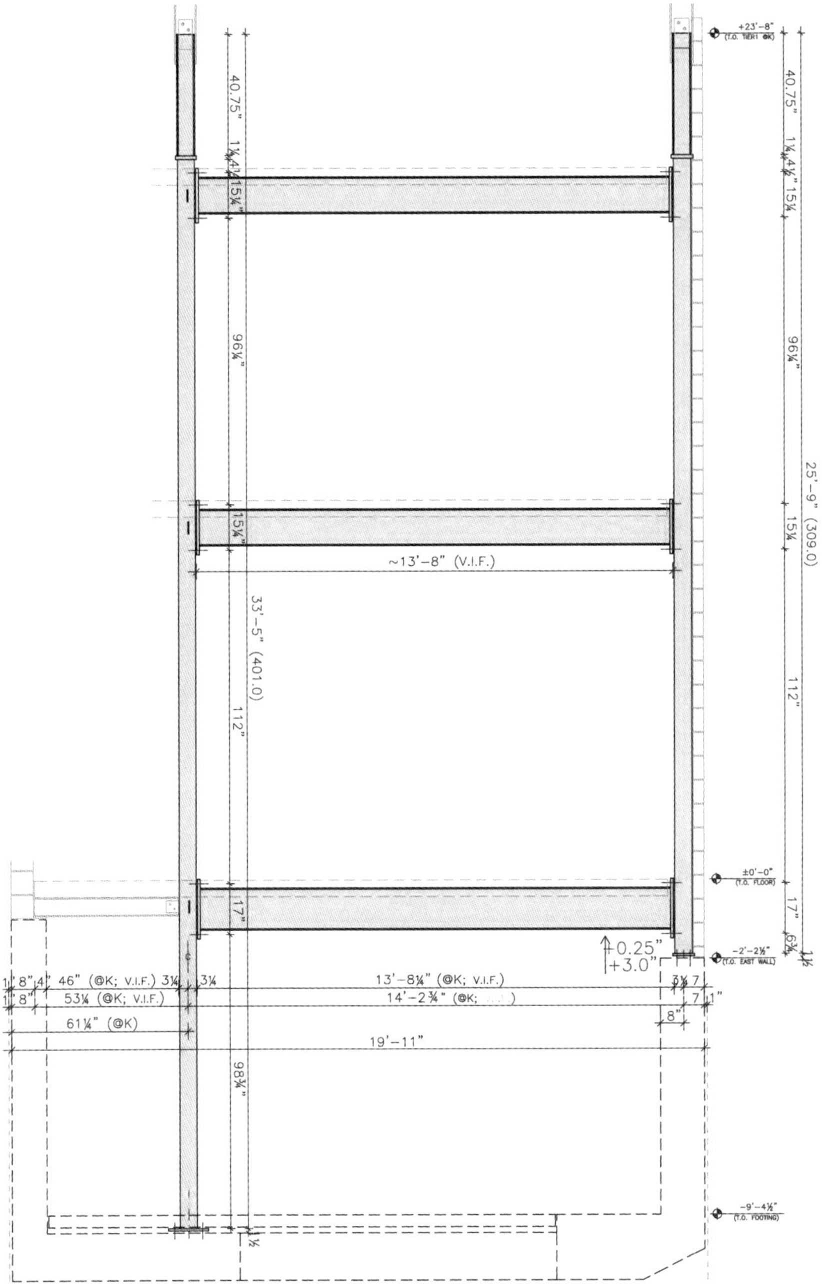

Verification of steel frame dimensions

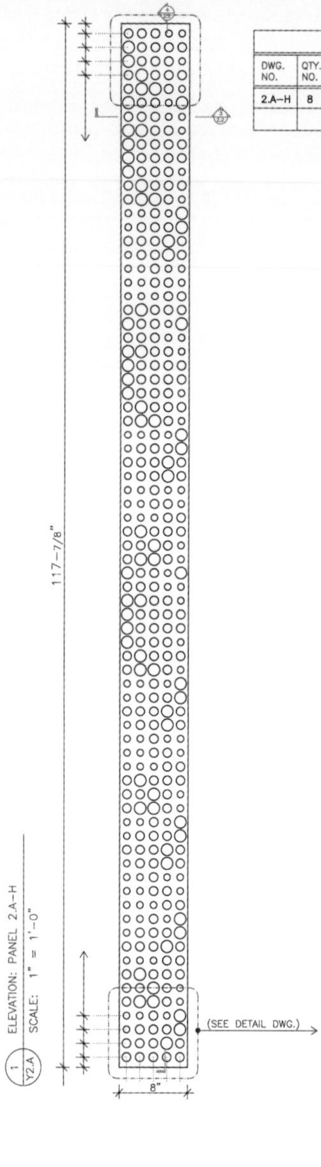

BILL OF MATERIAL						
DWG. NO.	QTY. NO.	SHIP MARK	TYPE	~FULL LENGTH	~FULL WIDTH	AREA [SF]
2.A-H	8	Y_2.A-H	PANELS	120"	~13"	10.83
						86.67

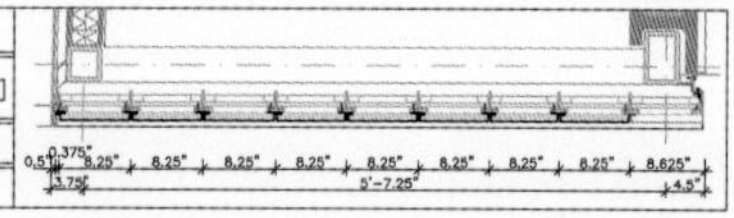

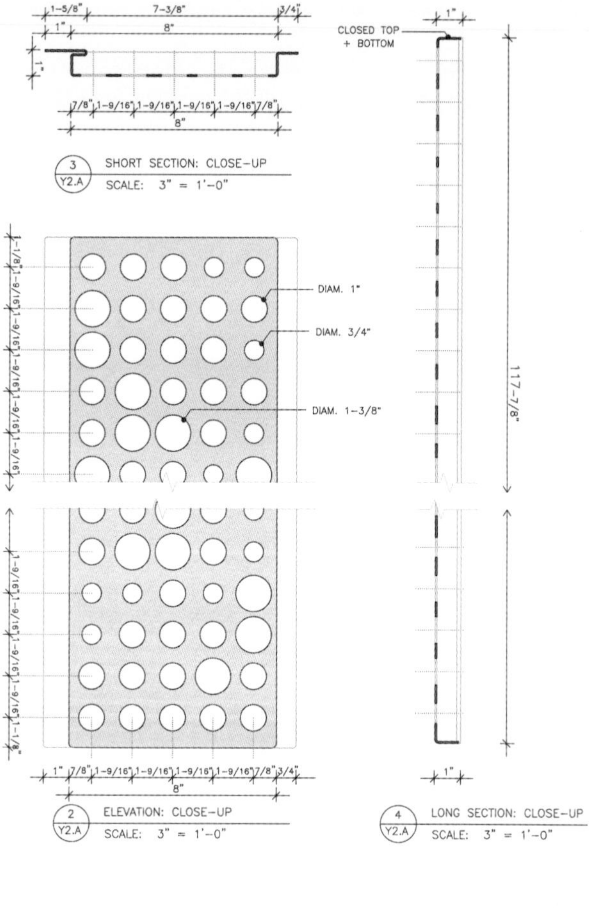

Perforated stair façade panel, detail study

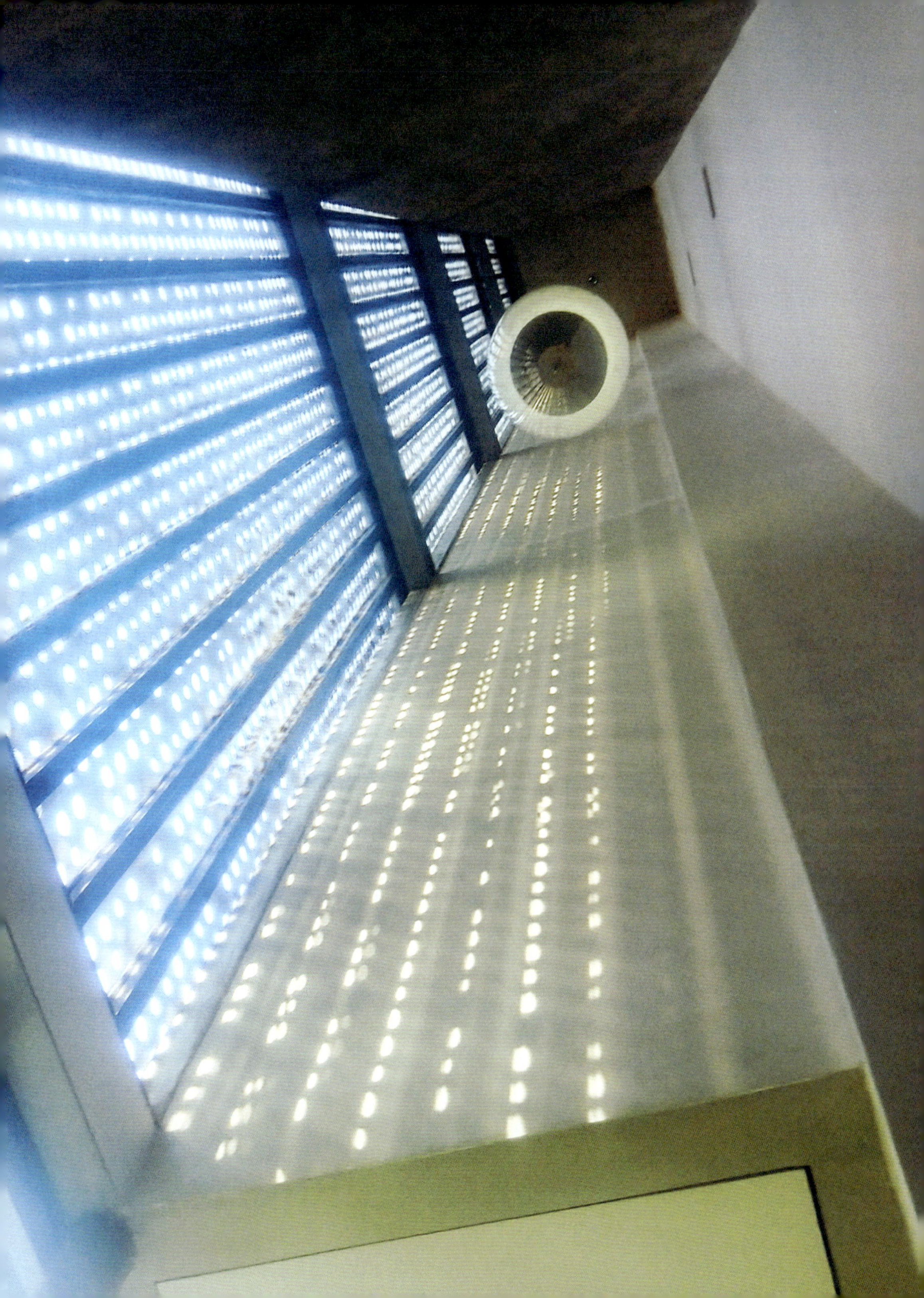

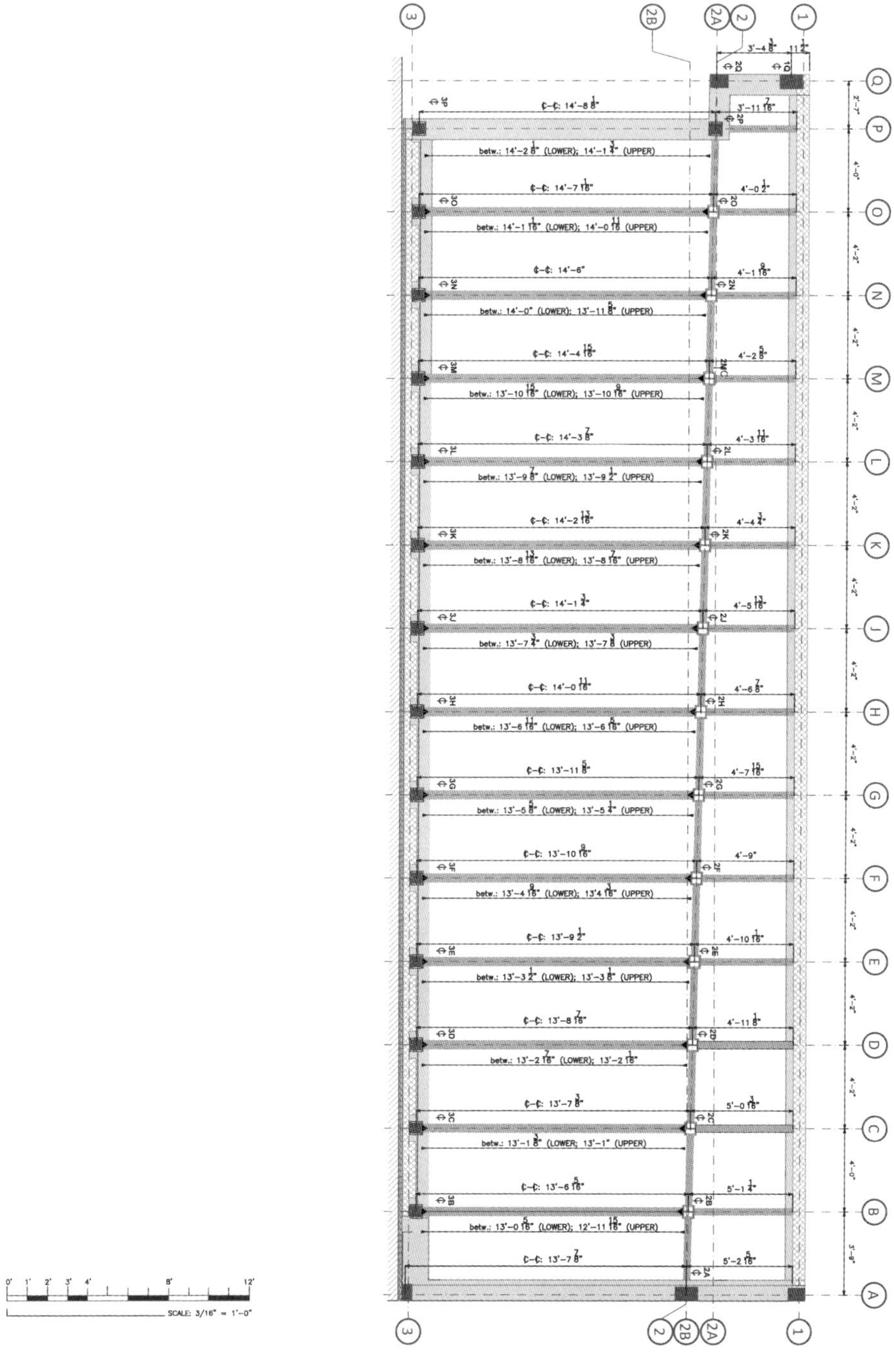

Steel Frame Geometry, Plan view

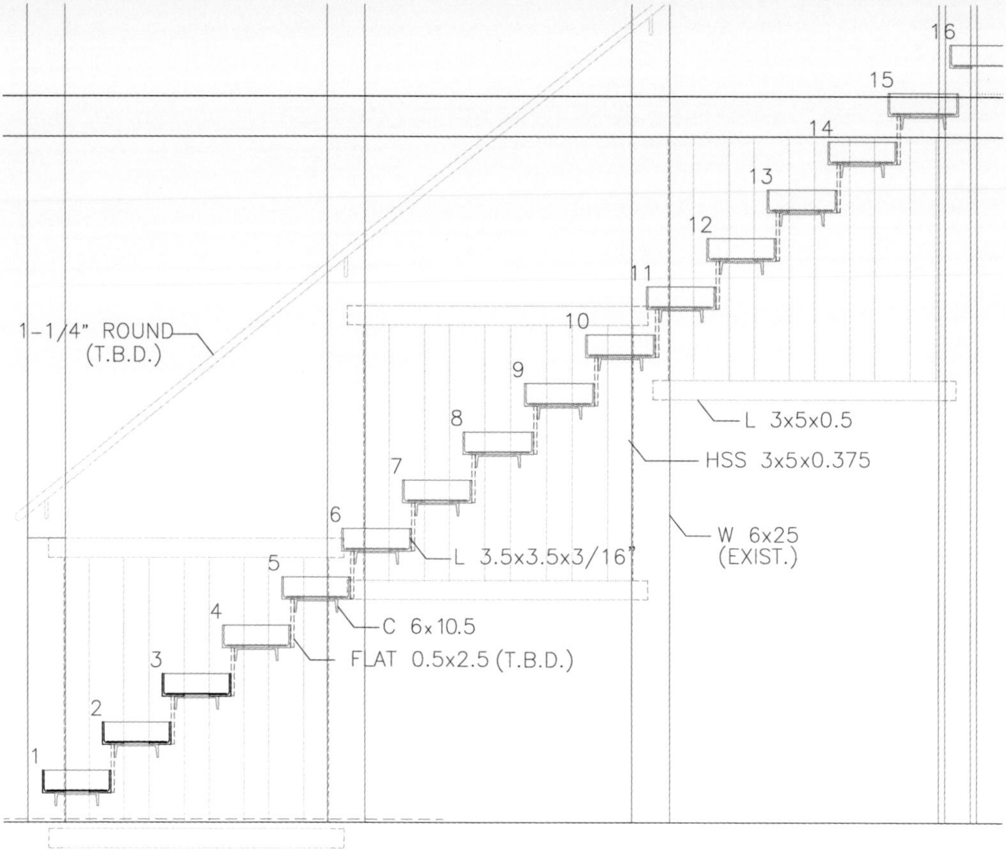

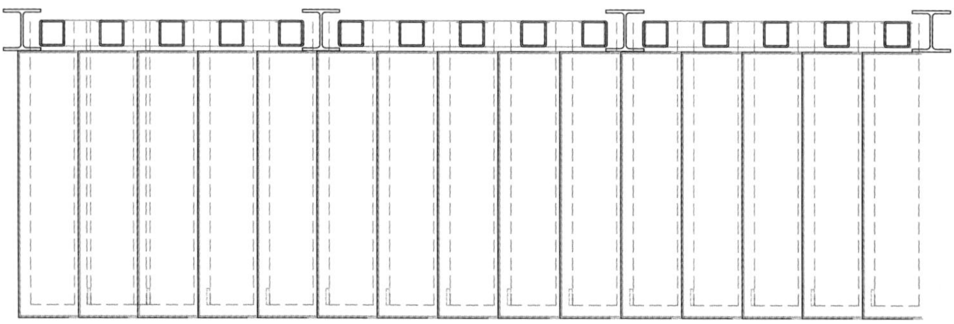

Stair to Roof, detail study

E

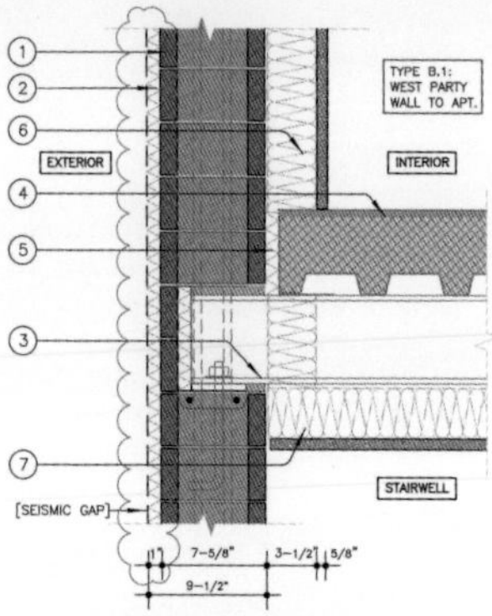

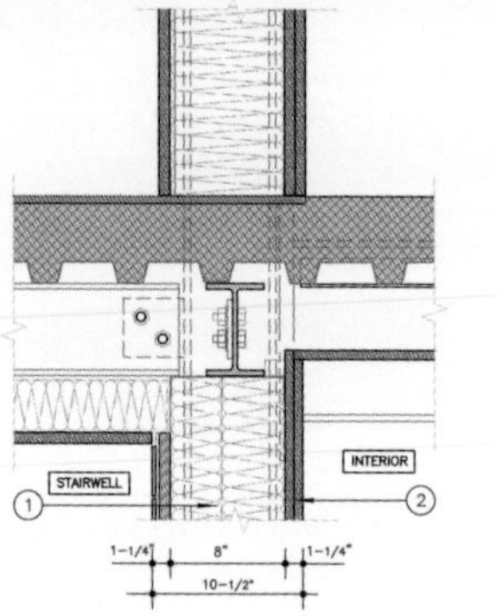

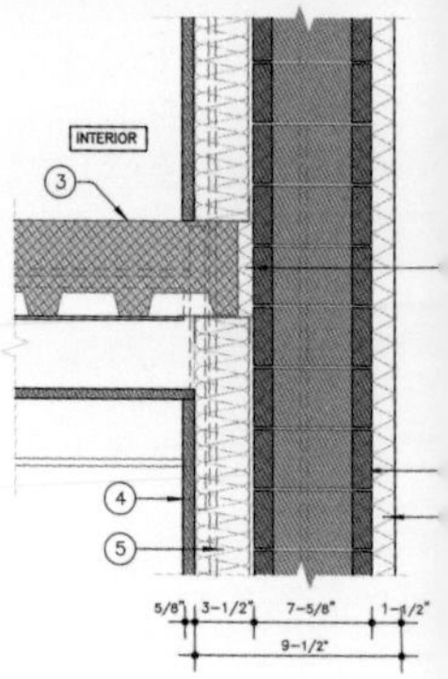

D

1. **CONCRETE MASONRY UNIT:** 7-5/8" WIDTH CMU BLOCK, PARTIALLY GROUTED CELLS — R-0.6

2. **RIGID INSULATION:** 1" SEISMIC GAP, POLYISOCYANURATE, R-6.5/INCH — [R-6.5]

3. **STEEL FRAME:** WIDE-FLANGE STEEL BEAM, A 36, ATTACHMENT: SEE STRUCTURAL DWGS. — N/A

4. **FLOOR SLAB:** CORRUGATED DECKING, 20 GA., 1-1/2" DEPTH — N/A

5. **RIGID INSULATION:** 1" AT CORRUGATED DECKING, R-5.5/INCH — [R-5.5]

6. **DRYWALL w/ BATT INSULATION:** 5/8" GWB [R-0.55] w/ 3.5" BATT INSULATION, OWENS-CORNING, R-3.5/INCH= R-19.25* — [R-12.8]
ALT.: INSULATION: OPEN-CELL FOAM INSULATION, SAME R-VALUE

7. **SUSPENDED CEILING:** 5/8" GWB [R-0.55] w/ 3.5" BATT INSULATION, OWENS-CORNING, R-3.5/INCH= R-12.25* — N/A

TYPE A.2: STAIR WALL, WEST (EXCL. EXTRA INSULATION TO APT.) — R-7.1 / U-0.14

TYPE A.1: FLOOR SLAB @ CMU DEMISING WALL, WEST — R-12.6 / U-0.08

TYPE B.1: DEMISING WALL, WEST (@ APARTMENTS) — R-19.9 / U-0.05

1/EN TYPE A.1: FLOOR SLAB/TYPE A.2: STAIR WALL/TYPE B.1
SCALE: 1-1/2"= 1'-0"

FROM DWG #1: TYPE A.2: STAIR WALL ASSEMBLY: — R-0.6

1. **BATT INSULATION:** 8" BATT INSULATION, OWENS-CORNING, R-3.5/INCH, CONT. ACOUSTIC BARRIER — R-28.0
ALT.: INSULATION: OPEN-CELL FOAM INSULATION, SAME R-VALUE

2. **DRYWALL:** 3x 5/8" DRYWALL ON 8" LIGHT-GAUGE FRAMING, CLARK DIETRICH , OR SIM.; R-0.56/BOARD — R-1.7

TYPE A.2: STAIR WALL, INCL. PARTITION WALL TO APT. — R-30.3 / U-0.033

2/EN TYPE A.2: STAIR WALL INCL. INT. WALL INSULATION
SCALE: 1-1/2"= 1'-0"

1. **CONCRETE MASONRY UNIT:** 7-5/8" WIDTH CMU BLOCK, GROUTED CELLS

2. **RIGID INSULATION:** 1.5" THICK POLYSTYRENE (XPS), 24" W w/ MECH. FASTENERS, TAPED; R-5.5/IN

3. **FLOOR SLAB:** CORRUGATED DECKING, 20 GA., 1-1/2" DEPT

4. **DRYWALL:** 5/8", ON LIGHT-GAUGE FRAMING, 16 O.C., TAPE

5. **BATT INSULATION:** 3.5" BATT INSULATION, OWENS-CORNING, R-3.5/INCH
ALT.: INSULATION: OPEN-CELL FOAM INSULATION, SAME R

6. **RIGID INSULATION:** 1" AT CORRUGATED DECKING, R-5.5/IN

TYPE A.3: FLOOR SLAB @ CMU PARTY WALL, EAST — R

TYPE B.2: DEMISING WALL, EAST

3/EN TYPE A.3: FLOOR SLAB / TYPE B.2: D
SCALE: 1-1/2"= 1'-0"

C

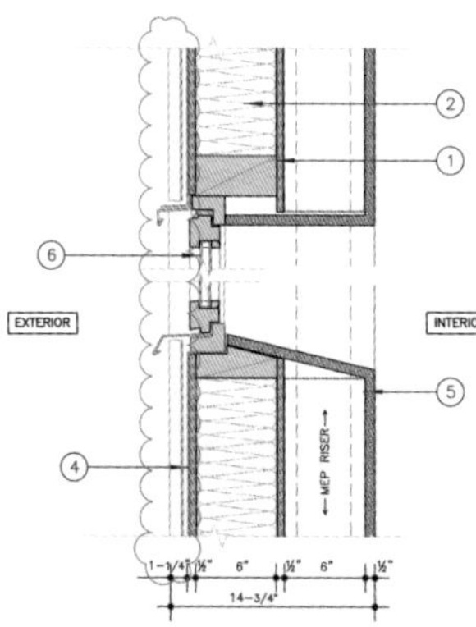

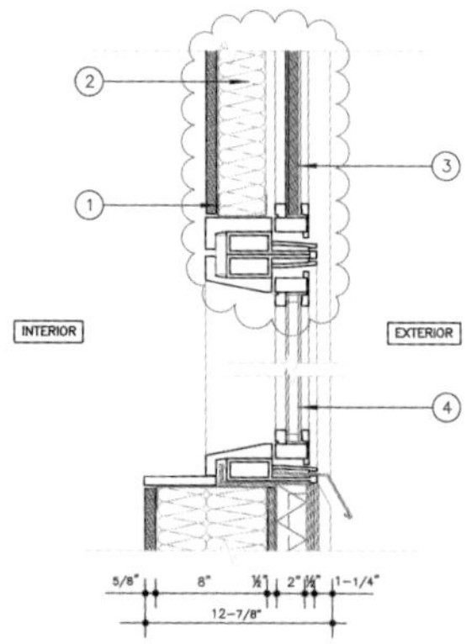

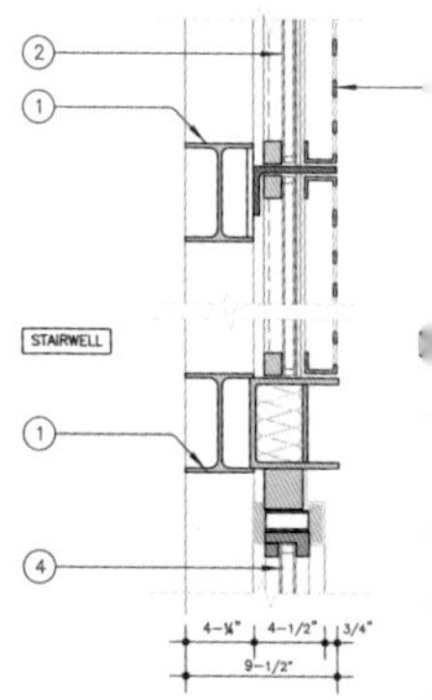

B

1. **LIGHT-GAUGE FRAMING:** 6" WIDTH, w/ 1/2" DENSE GLASS SHEATHING, CLARK-DIETRICH — R-0.55

2. **BATT INSULATION:** 6" THICK CAVITY INSULATION, OWENS-CORNING, R-3.5/INCH — R-21.0
ALT.: INSULATION: OPEN-CELL FOAM INSULATION, SAME R-VALUE

3. ~~**RIGID INSULATION:** 2" AS PART OF PROPRIETARY RAINSCREEN ASSEMBLY SYSTEM, KME/TECU, R-5.5/INCH — R-11.0~~

4. **METAL PANELING:** VMZinc PROPRIETARY SYSTEM, ON DENSE-GLASS, 1/2" w/ WATERPROOF MEMBRANE — R-0.55

5. **LIGHT-GAUGE FRAMING:** 6" WIDTH, w/ 1/2" DRYWALL INTERIOR, FINISH, CLARK DIETRICH; MEP RISER — N/A

6. **STAIR WINDOW:** TISCHLER (OR SIM.) w/ IGU, NFRC 100-2010, LOW-E GLASS, AVG. R-2.5, SHGC 0.60 — [R-2.5]

TYPE Q: STAIR REAR FACADE, ST. SEAM METAL PANELING — R-22.1 / U-0.05

TYPE M.1: STAIR WINDOW, 3.2 S.F. — R-2.5 / U-0.4

7/EN TYPE Q.1: STAIR, REAR FACADE /TYPE M.1: STAIR WINDOW
SCALE: 1-1/2"= 1'-0"

1. **LIGHT-GAUGE FRAMING:** 3-5/8" WIDTH, w/ 5/8" DRYWALL — R-0.55

2. **BATT INSULATION:** 3-1/2" THICK CAVITY INSULATION, OWENS-CORNING, R-3.5/INCH — R-12.25
ALT.: INSULATION: OPEN-CELL FOAM INSULATION, SAME R-VALUE

3. **SPANDREL PANEL:** THERMOLITE, 1" THICK COMPOSITE INSULATION PANEL — R- 3.9

4. **WINDOWS:** KOLBE 'VISTALUXE' CURTAIN WALL SYSTEM, w/ LoE-270 IGU w/ ARGON FILL, AVG. R-2.5, SHGC 0.60 — [R-2.5]

TYPE N: SPANDREL PANEL, REAR STAIR — R-16.7 / U-0.06

TYPE J.2: STAIR WINDOW REAR, EAST — R-2.5 / U-0.40

8/EN TYPE J.2: CURTAIN WALL / TYPE N: SPANDREL
SCALE: 1-1/2"= 1'-0"

1. **STEEL FRAME:** WIDE-FLANGE STEEL BEAM, A 36, SEE STRUCTURAL DWGS.

2. **INSULATED GLASS UNIT:** 1" THICK, NFRC 100-2010, AVG. R-2.6, SHGC 0.20

3. **PERFORATED METAL PANELS:** VMZinc PROP. SYSTEM SCREWED TO STEEL SUB-STRUC

4. **ENTRY STORE FRONT:** SCHÜCO PROP. SYSTEM, w/ 1" IGU, CLEAR GLASS, AVG. R-2.6

TYPE P.1: STAIR FACADE, NORTH

TYPE M.2: STORE FRONT ENTRY

9/EN TYPE P.1: STAIR NORTH/TYPE M.2:
SCALE: 1-1/2"= 1'-0"

A

Energy Compliance: detail studies

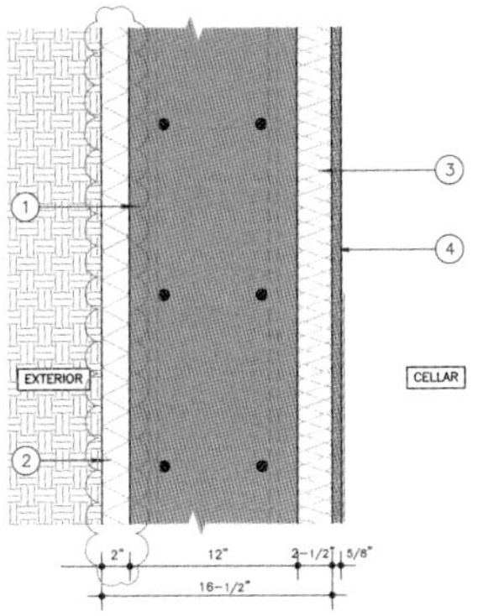

1. **CONCRETE**: 12" WIDTH REINFOCED CONCRETE — R-1.0
2. **RIGID INSULATION**: 2" THICK POLYSTYRENE (XPS), 24" WIDE, FULLY TAPED AT SEAMS, R-5.5/INCH — R-11
3. **BATT INSULATION**: 2.5" BATT INSULATION, OWENS-CORNING, R-3.5/INCH — R-8.8
4. **DRYWALL**: 5/8", ON LIGHT-GAUGE FRAMING, 16 O.C., TAPED — R-0.6

TYPE E.1: CELLAR WALL — R-21.4 / U-0.05

(4/EN) TYPE E.1: CELLAR WALL
SCALE: 1-1/2"= 1'-0"

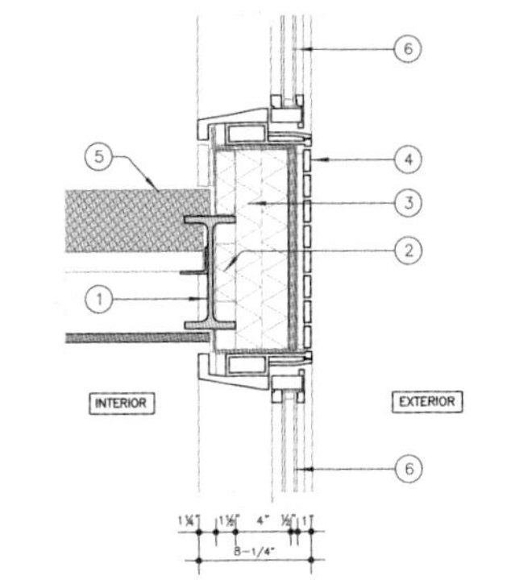

1. **STEEL FRAME**: WIDE-FLANGE STEEL BEAM, A 36, ATTACHMENT: SEE STRUCTURAL DWGS — N/A
2. **RIGID INSULATION**: 1.5" THICK POLYSTYRENE (XPS), WITHIN STEEL FLANGE CAVITY, R-5.5/INCH — R-8.3
3. **RIGID INSULATION**: 4" THICK POLYSTYRENE (XPS) w/ LIGHT-GAUGE FRAMING, R-5.5/INCH — R-22.0
4. **RAIN SCREEN**: TRESPA PROPRIETARY SYSTEM, ON DENSE GLASS, 1/2" w/ WATERPROOF MEMBRANE — N/A
5. **FLOOR SLAB**: CORRUGATED DECKING, 20 GA., 1-1/2" DEPTH — N/A
6. **WINDOWS**: RAICO 'THERM+' CURTAIN WALL SYSTEM, w/ LowE IGU w/ ARGON FILL, AVG. R-4.0, SHGC 0.60 — [R-4.0]

TYPE H.1: SPANDREL @ NORTH FACADE — R-30.3 / U-0.03

TYPE J.1: CURTAIN WALL @ NORTH FACADE — R-4.0 / U-0.25

(5/EN) TYPE H.1: SPANDREL / TYPE J.1: CURTAIN WALL
SCALE: 1-1/2"= 1'-0"

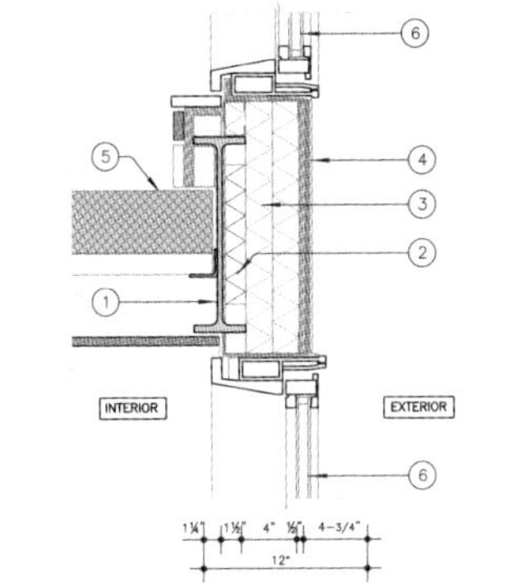

1. **STEEL FRAME**: WIDE-FLANGE STEEL BEAM, A 36, ATTACHMENT: SEE STRUCTURAL DWGS — N/A
2. **RIGID INSULATION**: 1.5" THICK POLYSTYRENE (XPS), WITHIN STEEL FLANGE CAVITY, R-5.5/INCH — R-8.3
3. **RIGID INSULATION**: 4" THICK POLYSTYRENE (XPS) w/ LIGHT-GAUGE FRAMING, R-5.5/INCH — R-22.0
4. **SPANDREL**: ALPOLIC FR-PANEL, 4mm INTEGRATED IN CURTAIN WALL SYSTEM — N/A
5. **FLOOR SLAB**: CORRUGATED DECKING, 20 GA., 1-1/2" DEPTH — N/A
6. **WINDOWS**: RAICO 'THERM+' CURTAIN WALL SYSTEM, w/ LowE IGU w/ ARGON FILL, AVG. R-4.0, SHGC 0.60 — [R-4.0]

TYPE H.1: SPANDREL @ NORTH FACADE — R-30.3 / U-0.03

TYPE J.1: CURTAIN WALL @ NORTH FACADE — R-4.0 / U-0.25

(6/EN) TYPE H.2: SPANDREL / TYPE J.1: CURTAIN WALL
SCALE: 1-1/2"= 1'-0"

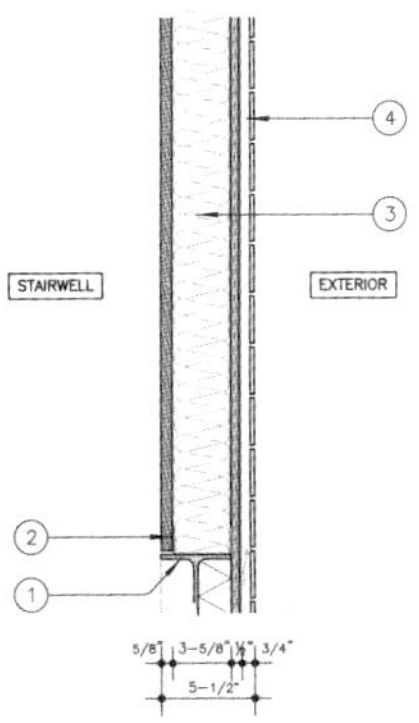

1. **STEEL FRAME**: WIDE-FLANGE STEEL BEAM, A 36, SEE STRUCTURAL DWGS — N/A
2. **LIGHT-GAUGE FRAMING**: 3-5/8" WIDTH, w/ 5/8" DRYWALL INT. FINISH, CLARK DIETRICH — R-0.55
3. **BATT INSULATION**: 3-1/2" THICK CAVITY INSULATION, OWENS-CORNING, R-3.5/INCH — R-12.25
   ALT.: INSULATION: OPEN-CELL FOAM INSULATION, SAME R-VALUE
4. **RAINSCREEN**: TRESPA PROPRIETARY SYSTEM, ON DENSE-GLASS, 1/2" w/ WATERPROOF MEMBRANE — R-0.55

TYPE O: STAIR FACADE, EAST — R-13.35 / U-0.07

(10/EN) TYPE O: STAIR EAST
SCALE: 1-1/2"= 1'-0"

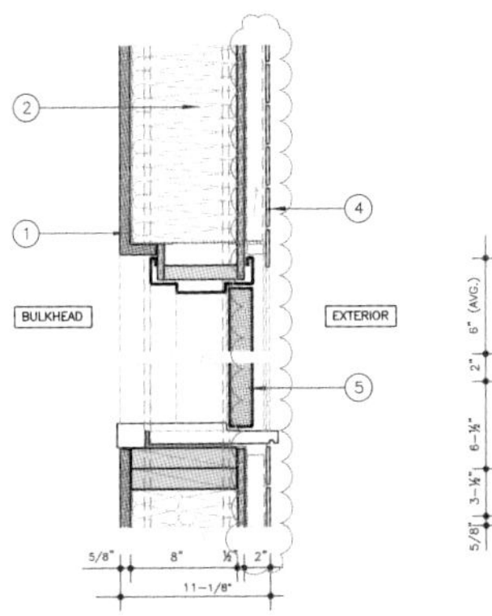

1. **LIGHT-GAUGE FRAMING**: 8" WIDTH, w/ 1/2" DENSE GLASS SHEATHING, CLARK-DIETRICH — R-0.55
2. **BATT INSULATION**: 6" THICK CAVITY INSULATION, OWENS-CORNING, R-3.5/INCH — R-21.0
   ALT.: INSULATION: OPEN-CELL FOAM INSULATION, SAME R-VALUE
3. ~~**RIGID INSULATION**: 2" AS PART OF PROPR. RAINSCREEN SYSTEM, TRESPA, R-5.5/INCH — R-11.0~~
4. **RAINSCREEN**: TRESPA PROPRIETARY SYSTEM, ON DENSE-GLASS, 1/2" w/ WATERPROOF MEMBRANE — R-0.55
5. **BULKHEAD DOOR**: 1-3/4" THICK, w/ POLYSTYRENE CORE, STEELCRAFT L-SERIES, R-1.1 — [R-1.1]

TYPE Q.2: BULKHEAD WALL, TRESPA RAINSCREEN — R-22.1 / U-0.05

TYPE L: BULKHEAD DOORS, 60 S.F. — R-1.1 / U-0.9

(11/EN) TYPE Q.2: BULKHEAD WALL / TYPE L: DOORS
SCALE: 1-1/2"= 1'-0"

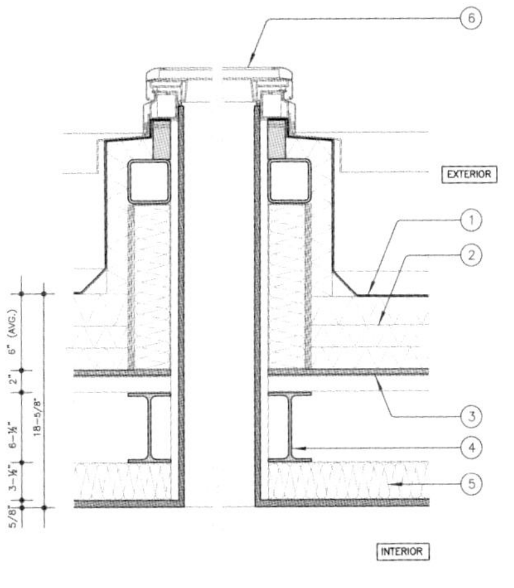

1. **ROOF MEMBRANE**: TPO 60 MIL., PITCHED TO DRAIN, JOHNS MANVILLE — N/A
2. **RIGID INSULATION**: 6" THICK AVG. INSULATION LAYER, DOW XPS, R-5.5/INCH — R-33.0
3. **ROOF DECK**: 20 GA. CORRUGATED DECKING — N/A
4. **STEEL FRAME**: WIDE-FLANGE STEEL BEAM, A 36, SEE STRUCTURAL DWGS. — N/A
5. **DROPPED CEILING**: 5/8" THICK DRYWALL, w/ 3-1/2" BATT INSULATION, R-3.5/INCH — R-12.8
6. **SKYLIGHT**: AUTOMATED OPERABLE SKYLIGHT w/ ARGON-FILLED LOW-E IGU, AVG. R-2.2. SHGC 0.60 — [R-2.2]

TYPE R: ROOF, TYPICAL — R-45.8 / U-0.02

TYPE S: SKYLIGHT, 14 S.F. — R-2.2 / U-0.45

(12/EN) TYPE R: ROOF, TYPICAL / TYPE S: SKYLIGHT
SCALE: 1-1/2"= 1'-0"

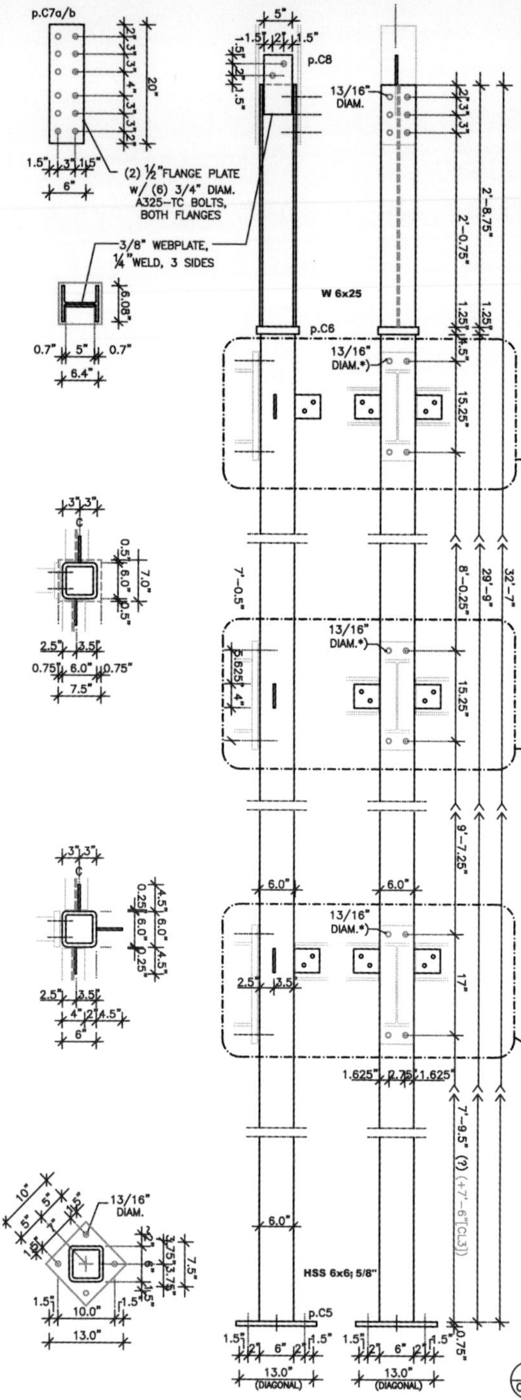

BILL OF MATERIAL

DWG. NO.	QTY. NO.	SHIP MARK	DESCRIPTION	LENGTH
	1	CL.C2	HSS 6x6, 5/8" TH.	29'-9"
	1	CL.C2	W 6x25	2'-8.75"
	1	p.C5	3/4" BASE PLATE, 10x10, A36	
	4		3/4" DIAM. F1554 GR. 36 ANCHOR RODS	
	1	p.C6	1-1/4" TRANS.PLATE, 7.5x7, A572 GR50	
	2	p.C7a/b	1/2"PLATE, 6x20", A36	
	1	p.C8	3/8"PLATE, 5x10", A36	
	26		3/4" DIAM. BOLTS, A325N	

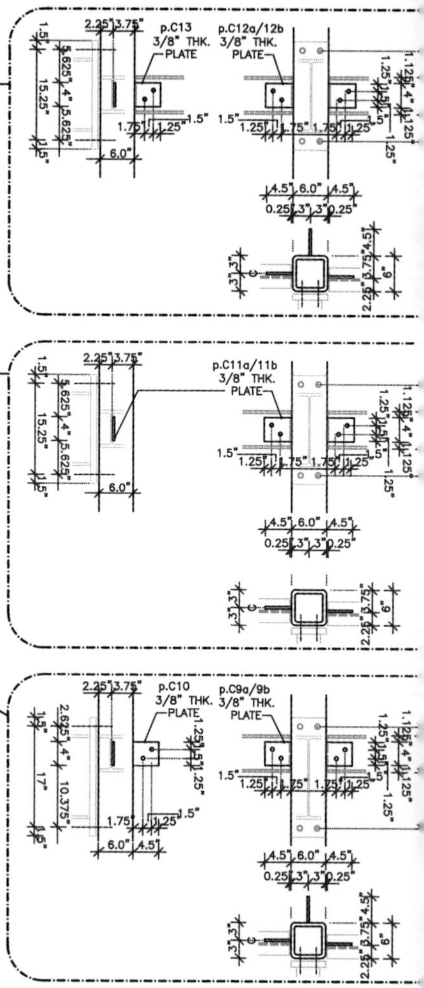

1 C.02 **FRONT, SIDE, TOP:** COLUMN, CENTERLINE C, LOWER TIER (CENTE

SCALE: 1/2" = 1'-0"

MATERIAL SPEC				
T	BOLT (UNO)	HOLES (UNO)	WELD	PAINT
	A325N (3/4")	(13/16")	(1/4")	ONE COAT: RUST INHIBITIVE PAINT
'p.C8				

NOTE:
ERECTOR/FABRICATOR SHALL VERIFY THAT BOLTS FIT AS PER MANF., OR SUBMIT REVISED DETAIL PRIOR TO DRILLING HOLES.

BILL OF MATERIAL				
TY. O.	SHIP MARK	DESCRIPTION	LENGTH	COMMENT
2	p.C12a/b	3/8"PLATE, 4x4.5", A36		
[illegible]	p.C13	3/8"PLATE, 4x4.5", A36		
[illegible]		3/4" DIAM. BOLTS, A325N,t.b.d.		for p.C12/p.C13

BILL OF MATERIAL				
TY. O.	SHIP MARK	DESCRIPTION	LENGTH	COMMENT
2	p.C11a/b	3/8"PLATE, 4x4.5", A36		
[illegible]		3/4" DIAM. BOLTS, A325N,t.b.d.		for p.C11a/b

BILL OF MATERIAL				
TY. O.	SHIP MARK	DESCRIPTION	LENGTH	COMMENT
2	p.C9a/b	3/8"PLATE, 4x4.5", A36		
[illegible]	p.C10	3/8"PLATE, 4x4.5", A36		
[illegible]		3/4" DIAM. BOLTS, A325N,t.b.d.		for p.C9/p.C10

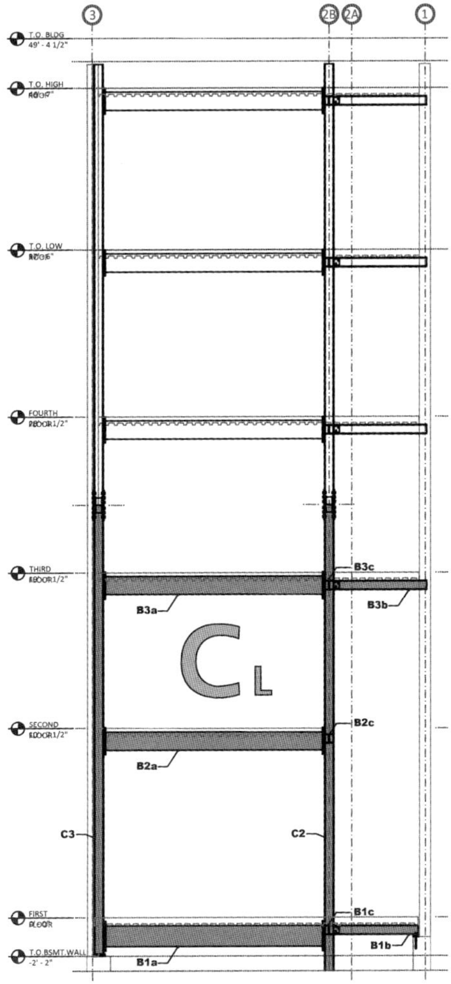

Steel column, shop drawing

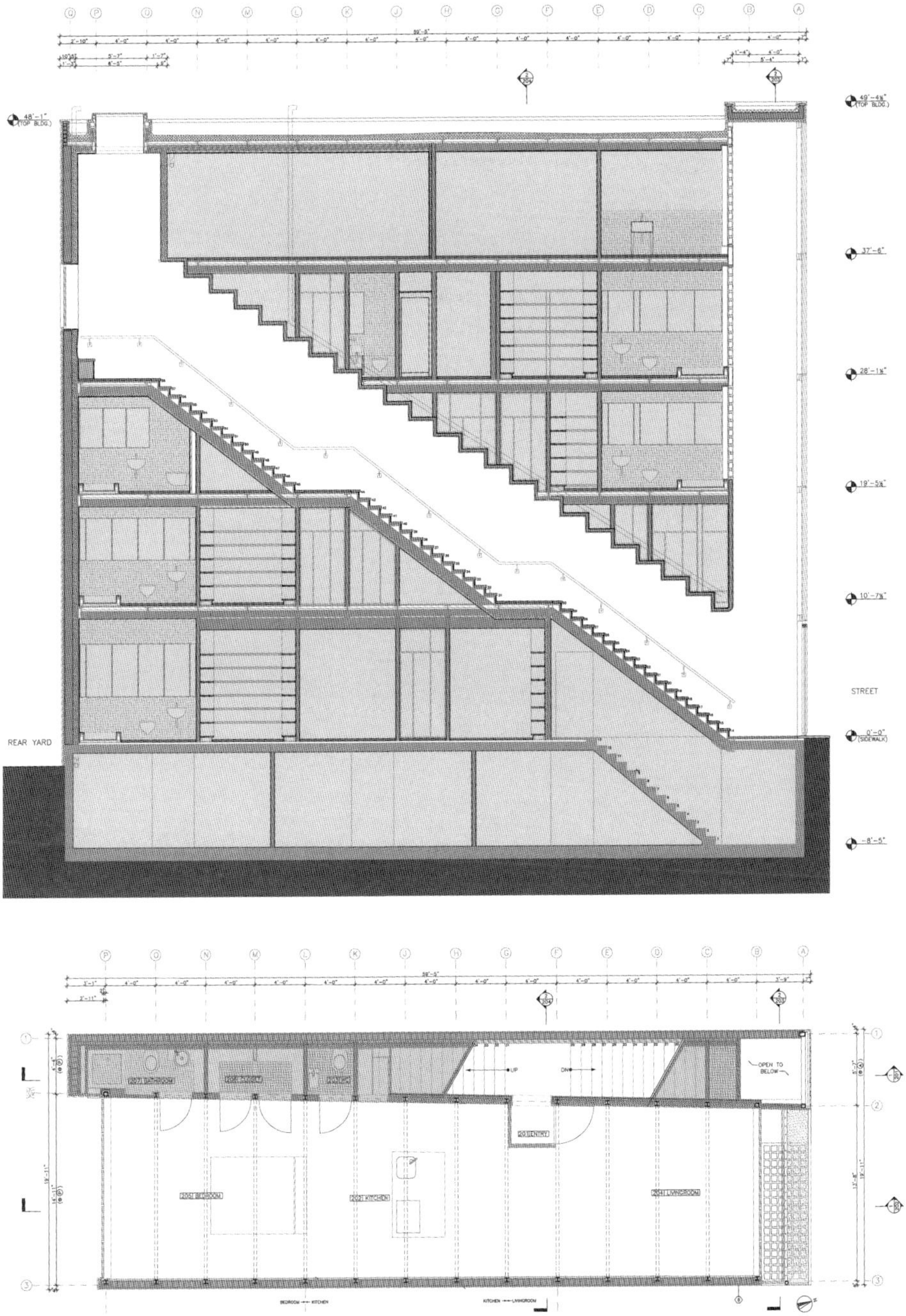

Section through Staircase + floor plan 2nd floor

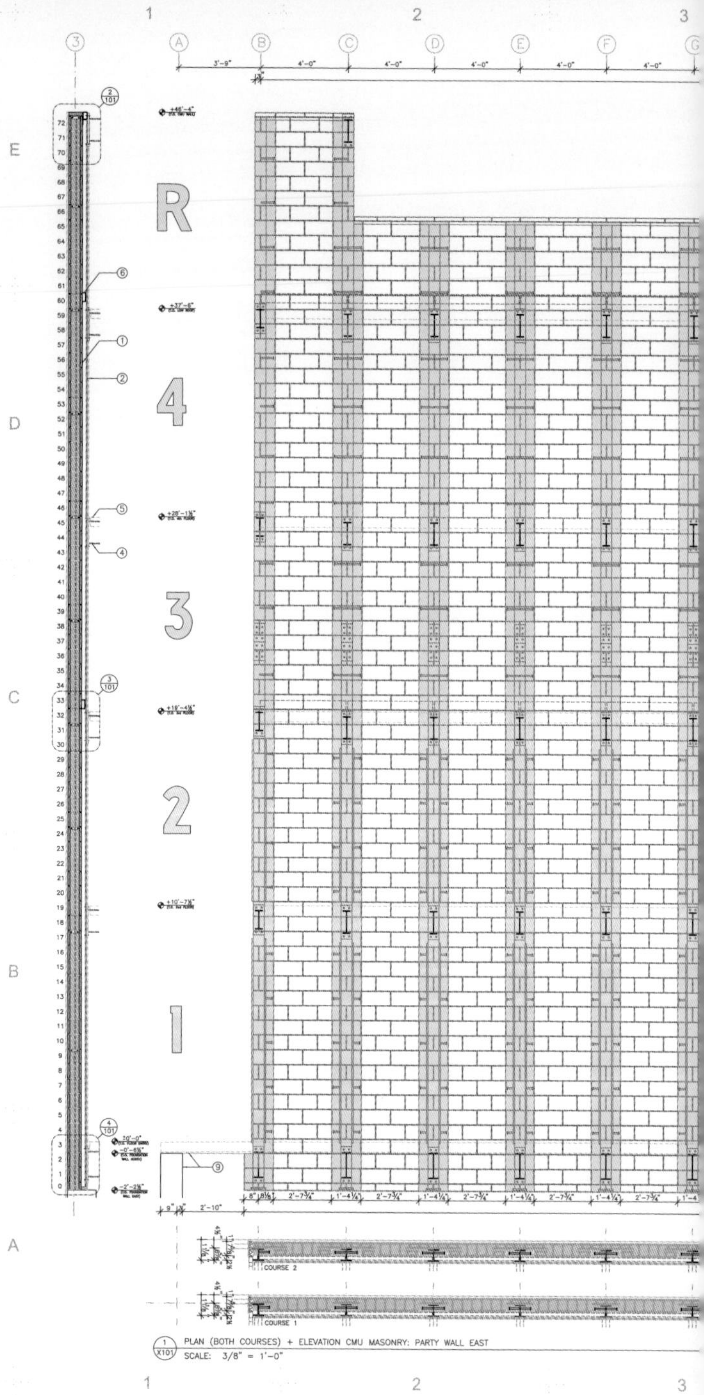

East demising wall, concrete masonry with embedded steel; elevation, section, plan + details

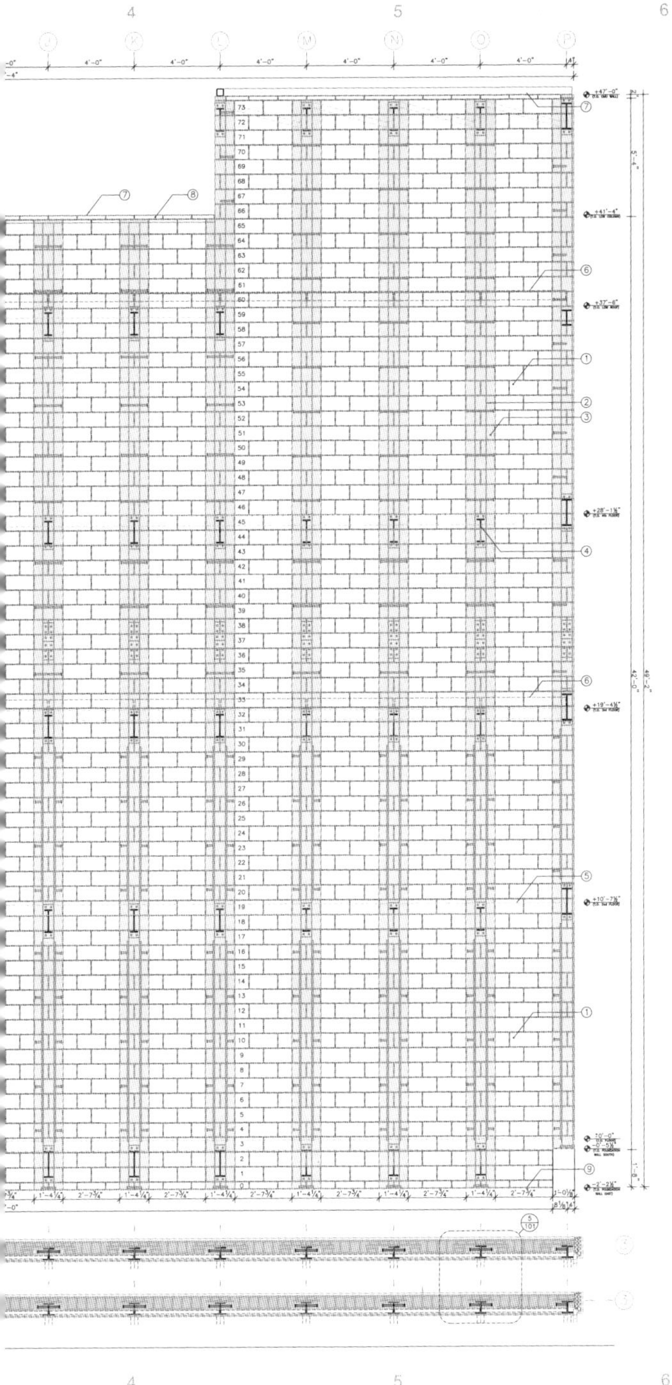

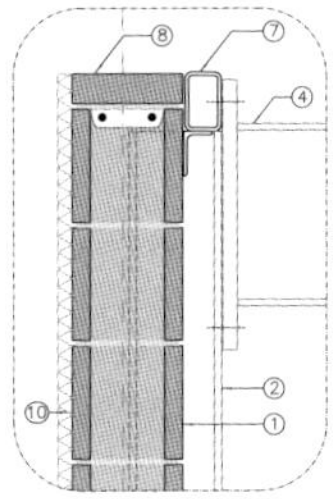

2 X101 DETAIL SECTION: TOP OF WALL
SCALE: 1-1/2" = 1'-0"

NOTE:
SEE STRUCTURAL DWG. SET FOR SPECIFIC REINFORCEMENT AND GROUTING.
SEE ALSO FOR ADDITIONAL DETAILS.

#	CMU TYPE [W]	CMU TYPE [L]	CMU TYPE [H]	DESCRIPTION	AMOUNT (UNITS)
01	8"	16"	8"	8" STANDARD	~1332
02	4"	16"	8"	4" STANDARD	~1352
03	4"	8"	8"	4" HALF	~ 896
04	8"	16"	4"	8" ½HGT. STAND.	~ 13
05	4"	16"	4"	4" ½HGT. STAND.	~ 40
06	8"	16"	2"	8" CAP, SOLID	~ 40

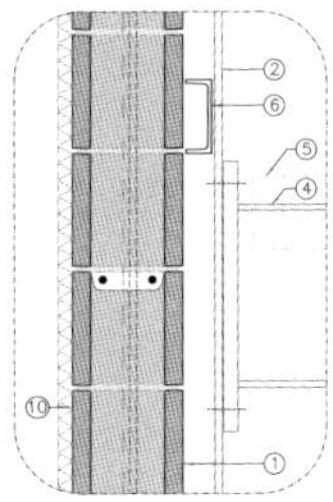

3 X101 DETAIL SECTION: MID-WALL
SCALE: 1-1/2" = 1'-0"

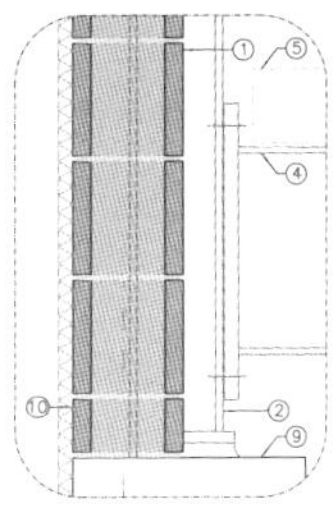

4 X101 DETAIL SECTION: BASE OF WALL
SCALE: 1-1/2" = 1'-0"

1. 8"/4" REINFORCED CMU WALL, 16" LENGTH, RUNNING BOND, SEE DETAILS + S-DRAWINGS
2. STEEL COLUMN, W6x25, FOR MOMENT FRAME STRUCTURE, w/ 7"/4" NELSON STUDS, 3/4" DIAM.
3. CMU POCKET, 4" DEEP, FOR EMBEDDED STEEL COLUMN, TO BE SOLIDLY GROUTED AFTER STEEL INSTALLATION, SEE STRUCT. DETAILS
4. STEEL BEAMS FOR MOMENT FRAMES, VARIOUS SIZES, BOLTED TO COLUMN
5. FINISHED FLOOR LEVEL, ON CORRUGATED DECKING, HELD BY STEEL BEAMS
6. STEEL SUPPORT CHANNELS, FOR COLUMN BRACING, C5x6.7 CHANNELS, BOLTED w/ HILTI SLEEVE ANCHORS, 32" O.C.
7. TOP OF STEEL COLUMN, HSS STEEL SECTIONS, HSS 4x4/4x2.5, 1/4" TH., T.B.D., SEE DETAIL
8. CMU CAP, 2", AT TOP OF REINFORCED CMU WALL, RUNNING BOND
9. FOUNDATION WALL, REINFORCED CONCRETE, 12"-16" THICK
10. SEISMIC GAP, DISTANCE TO NEIGHBORING PROPERTY, FILLED SUCCESSIVELY w/ RIGID INSULATION

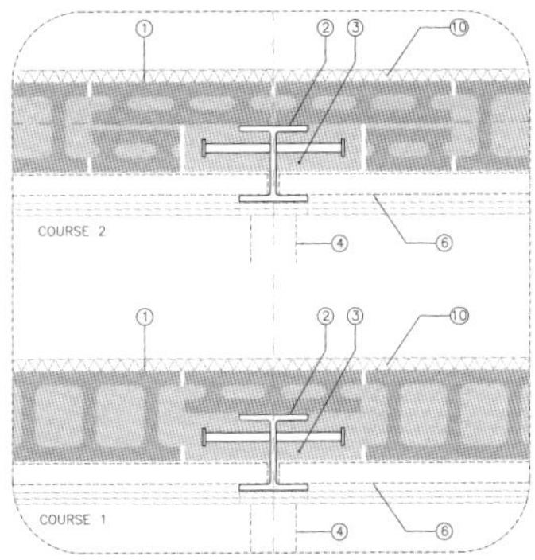

5 X101 DETAIL: PLAN VIEW (BOTH COURSES): AT STEEL MOMENT FRAME
SCALE: 1-1/2" = 1'-0"

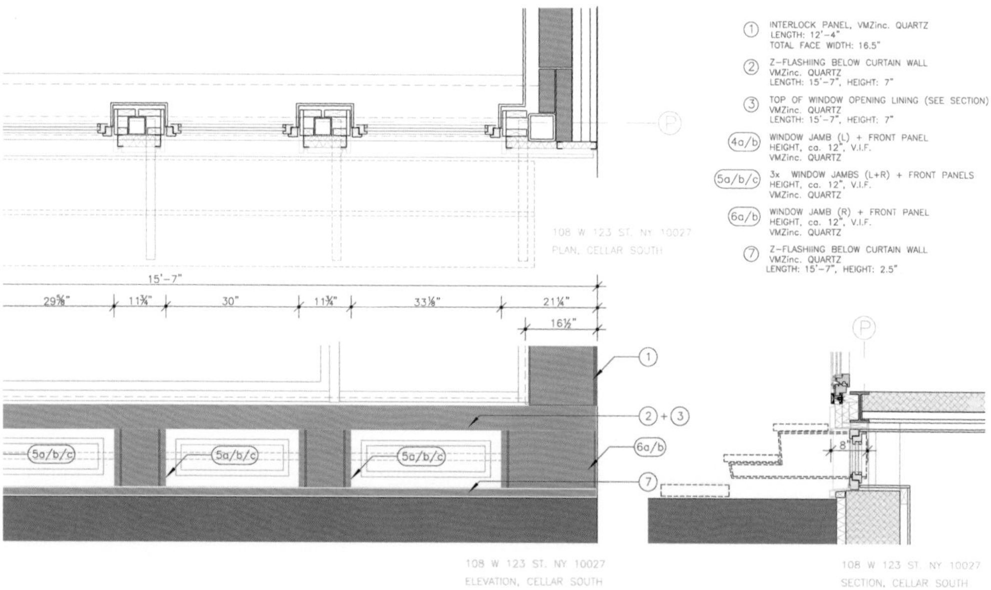

Floating stair and cellar windows, detail study

Site Safety planning: Sidewalk shed protection, elevation and section

5th STORY OF NEIGHB. BLDG., T.O.ROOF: 57'-3"
49'-6" (TOP BLDG.)
48'-2" (TOP ROOF)
37'-6" (TOP LOW ROOF)
28'-1½" (TOP 4th FL.)
19'-5½" (TOP 3rd FL.)
10'-7½" (TOP 2nd FL.)
0'-0" (SIDEWALK)
-8'-5" (TOP CELLAR)
CADE, FLUSH HB. BLDG.
8 FT. HGT. CONSTRUCTION FENCE
SHORING/LAGGING: 9-5/8" MINIPILES w/ WOOD LAGGING
9'-3"
KEY PLAN:
LENOX AVE.
W 123rd STREET
W 122nd STREET
A. CLAYTON POWELL BLVD.
NYC DEPARTMENT OF BUILDINGS
SITE SAFETY PLAN
ELEVATION SHEET, SUPERSTRUCTURE AT FRONT
DWG# SS-006.00
DATE: 11/15/18
SCALE: 3/16"=1'-0"
PAGE: 06 OF 09

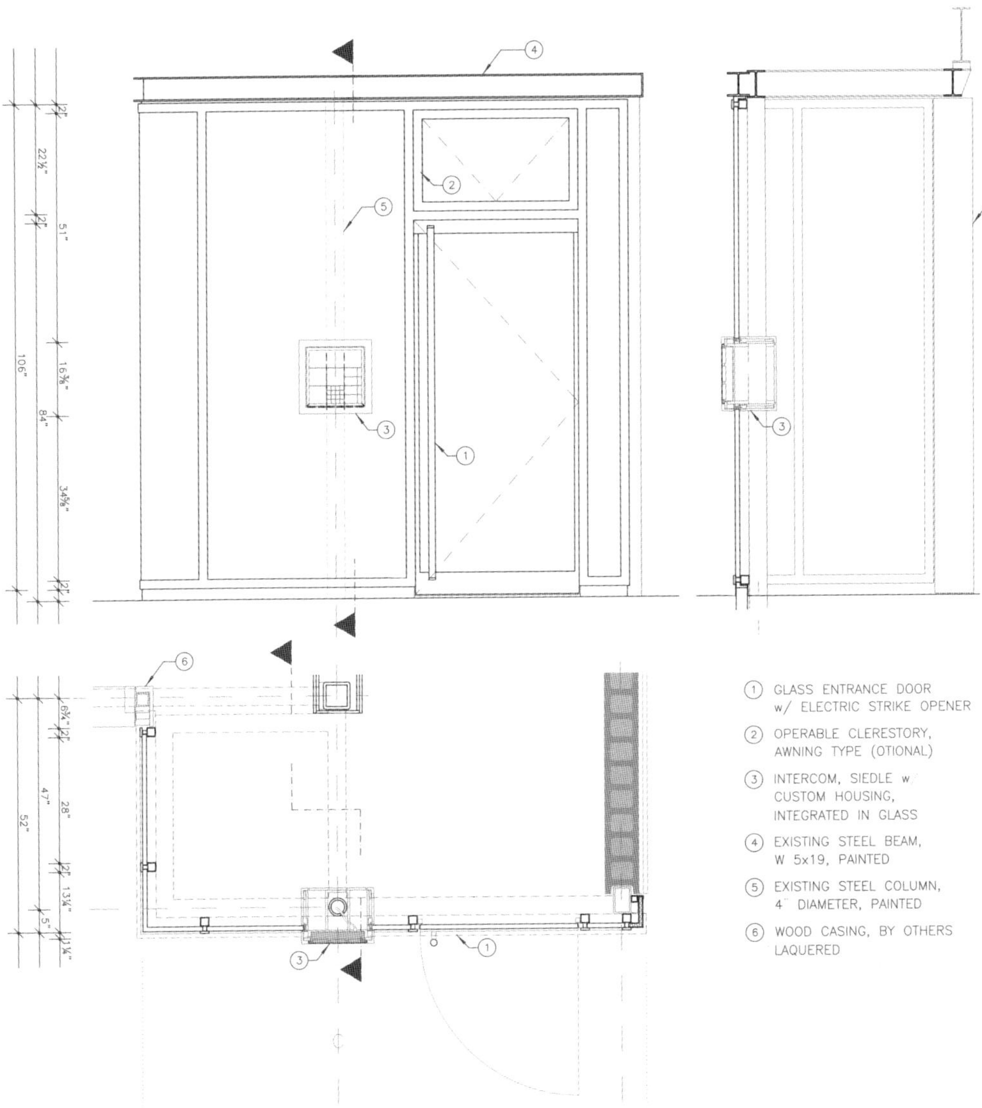

Entrance glazing and Intercom, plan, section + elevation

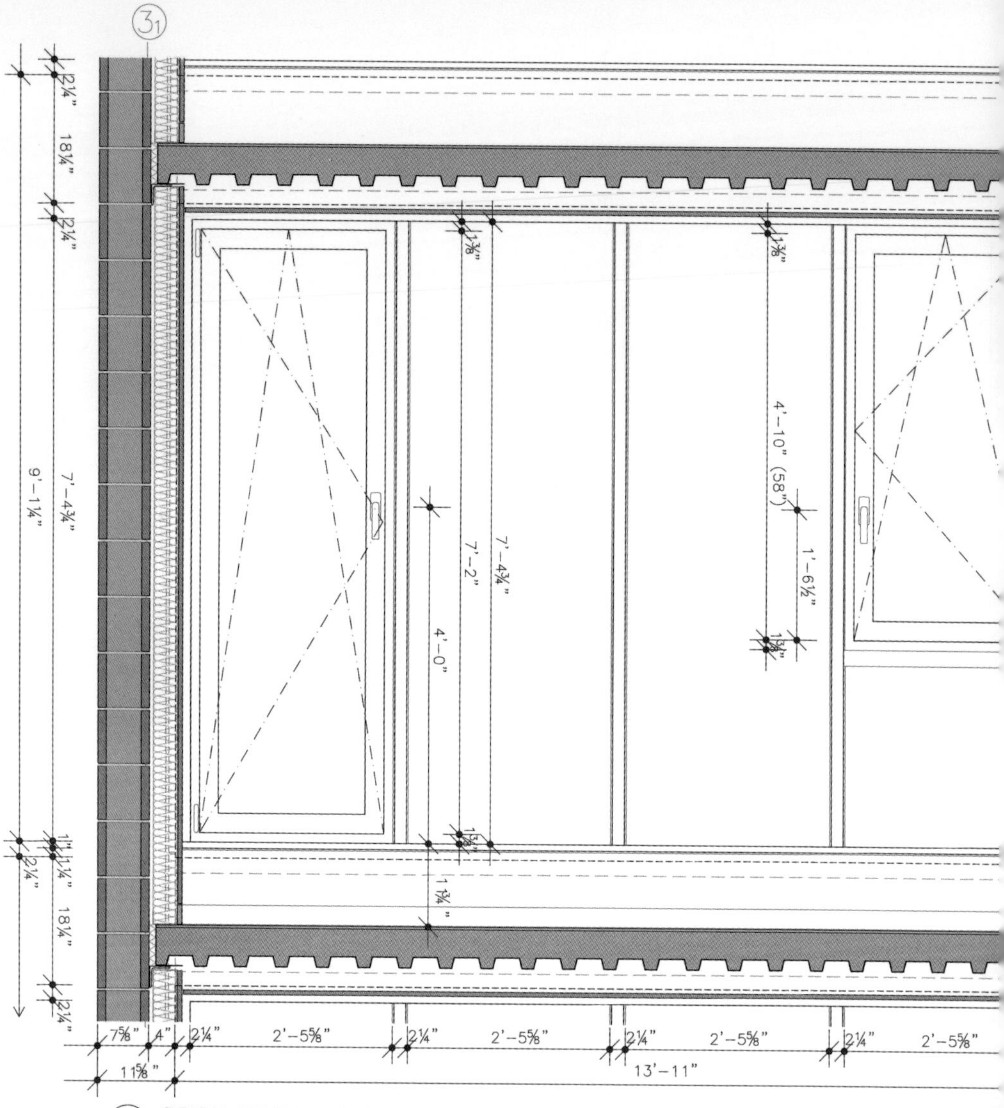

6
W2S
ELEVATION, WINDOW WALL SOUTH
SCALE: 3/8" = 1'–0"

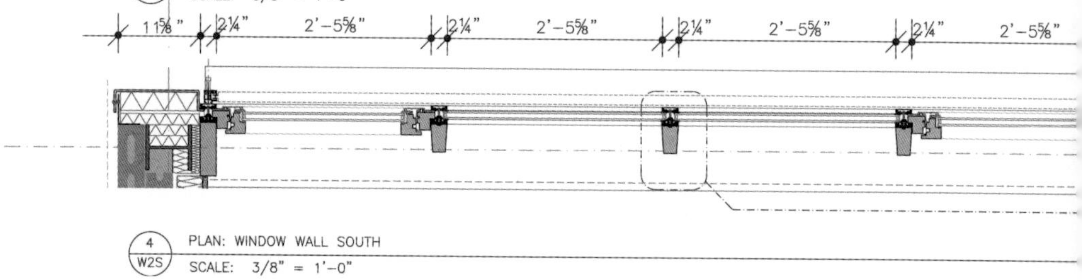

4
W2S
PLAN: WINDOW WALL SOUTH
SCALE: 3/8" = 1'–0"

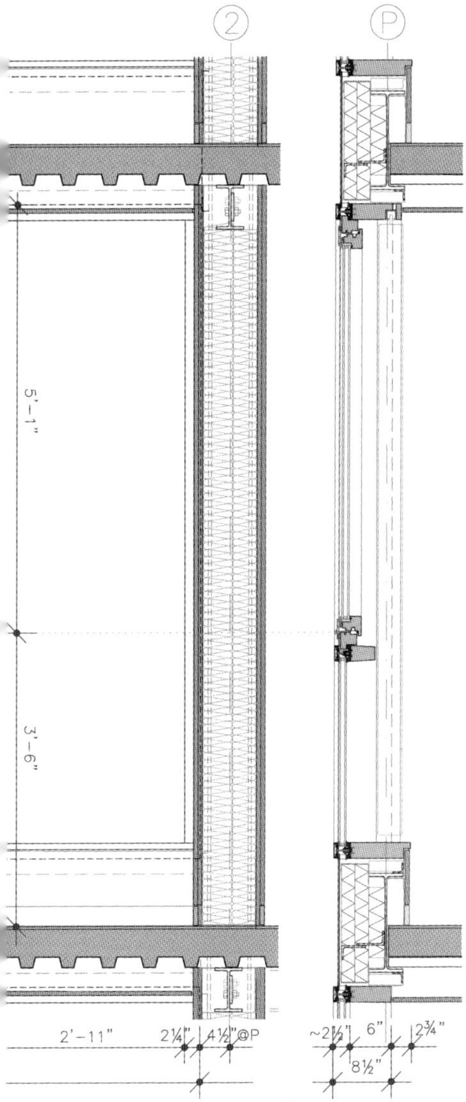

7 W2S SECTION, WINDOW WALL SOUTH
SCALE: 3/8" = 1'-0"

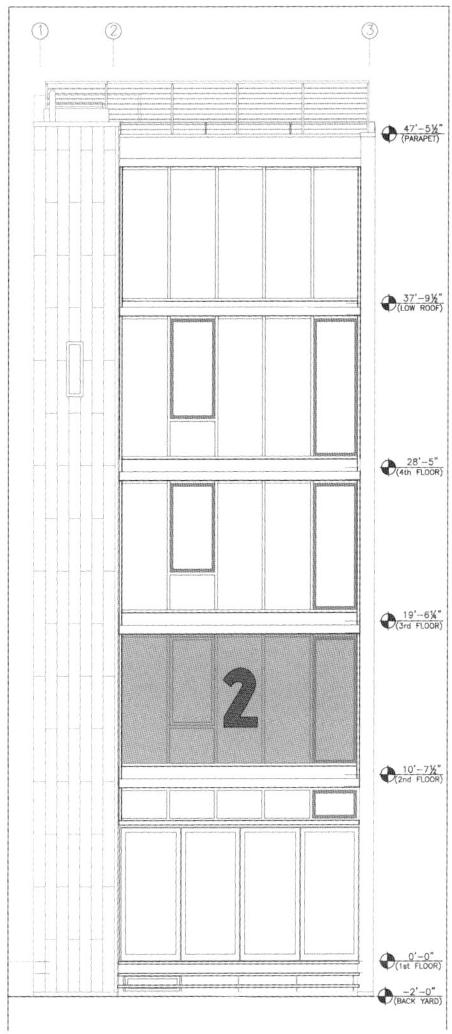

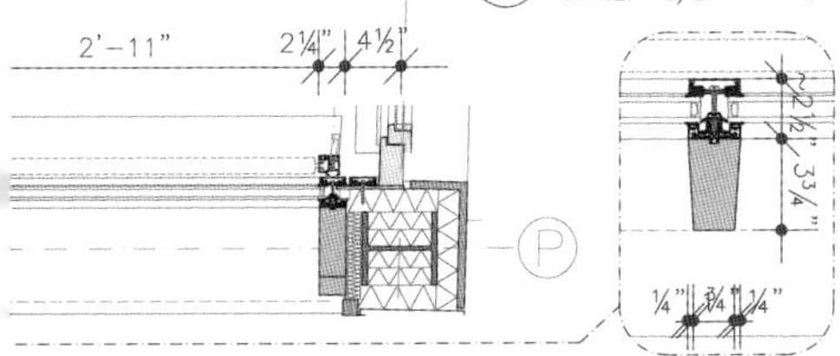

5 W2S DETAIL, CENTER POST TAPER
SCALE: 3/4" = 1'-0"

Window wall, main façade South, plan, section + elevation

APPROX. UNIT – TAKE-OFF						
#	[W]	[L]	[X]	DESCRIPTION	AMOUNT	ORDER
01	5"	3"	0.5"	ANGLE [5X3]	~ 14ft.	EXIST.
02	3"	3"	0.5"	ANGLE [3X3]	~ 43ft.	EXIST.
03	2"	2"	0.25"	TEE, 2"	~ 18ft.	ORDER
04	1"	X	0.25"	FLAT 1"	~ 8ft.	ORDER
05	32"	38"	.125"	SHEET, 1/8"	~25 sf	ORDER
06	1.25"	X	"	ST.STEEL ROUND	~ 13ft.	ORDER
07	0.75"	X	"	ST.STEEL ROUND	~ 4ft.	ORDER
08	2"	2"	"	SQUARE	~ 16ft.	ORDER
09	5"	X	0.25"	FLAT 1"	~ 13ft.	ORDER
NOTE: TAKEOFF MUST BE CONFIRMED BASED ON FINAL DESIGN DECISIONS (T.B.D.)						

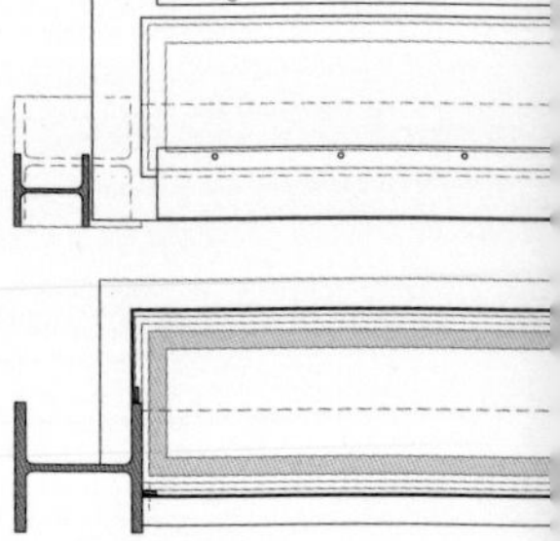

1 A201 PLAN + HORIZONTAL SEC

SCALE: 1/4" = 1'-0"

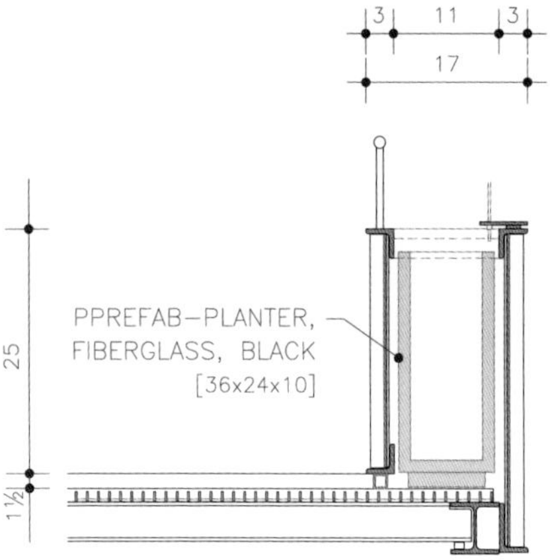

2 A201 LATERAL SECTION

SCALE: 1/4" = 1'-0"

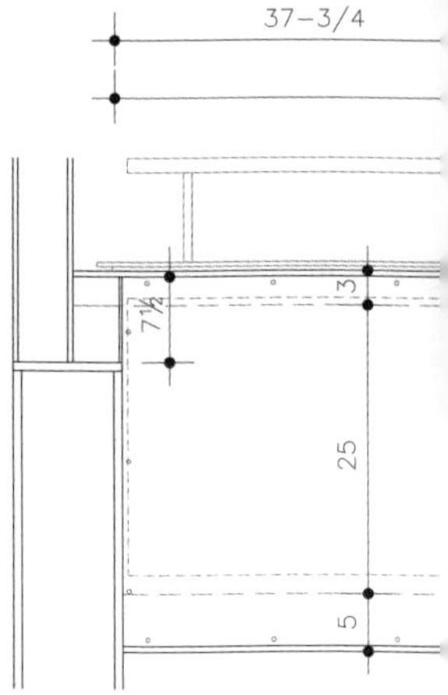

3 A201 ELEVATION: PLAMTER

SCALE: 1/4" = 1'-0"

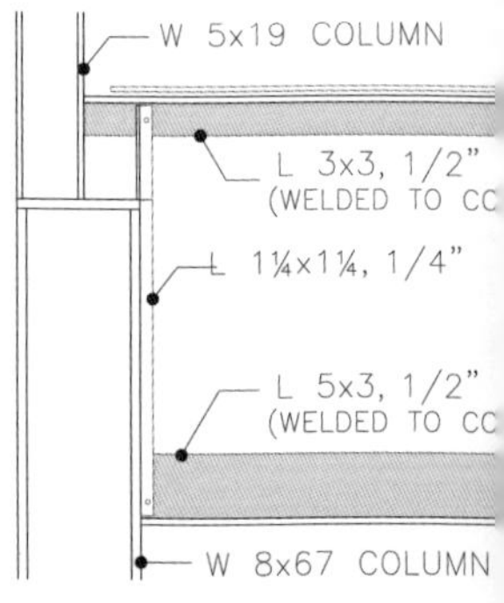

4 A201 ELEVATION FRAMING: PLAN

SCALE: 1/4" = 1'-0"

Main planter, street façade 2nd fl., plan, sections + elevation

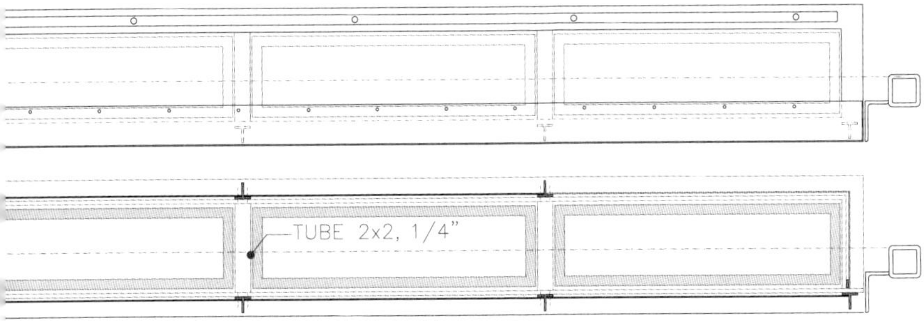
TUBE 2x2, 1/4"

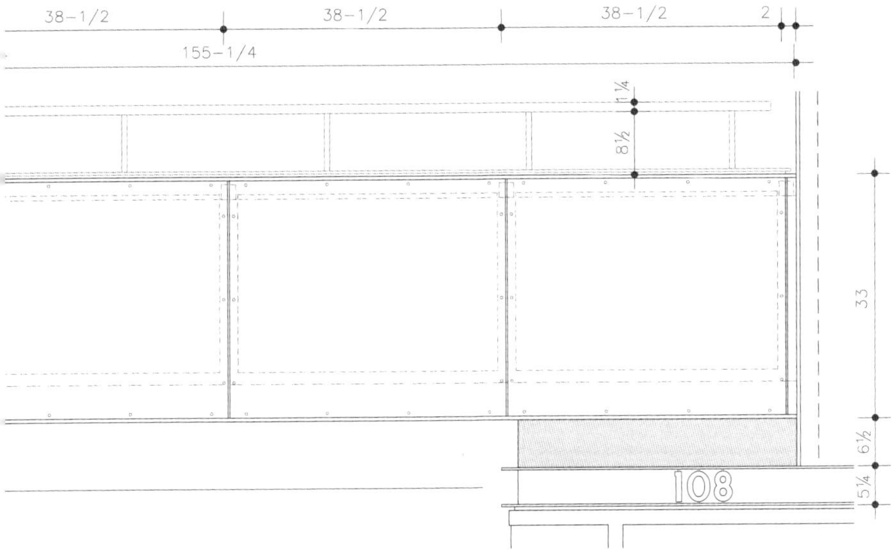
38-1/2
38-1/2
38-1/2
2
155-1/4
1¼
8½
33
6½
5¼
108

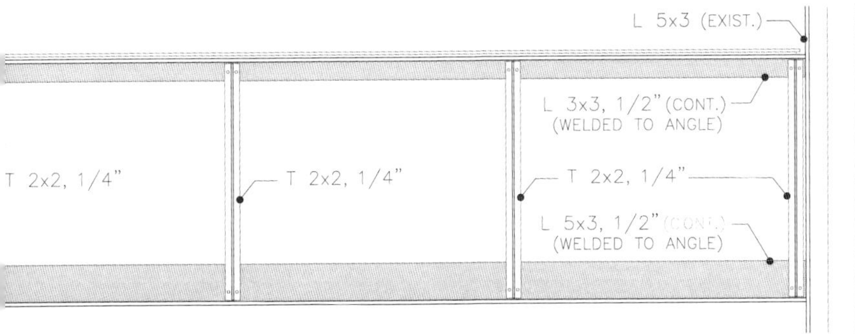
L 5x3 (EXIST.)
L 3x3, 1/2"(CONT.)
(WELDED TO ANGLE)
T 2x2, 1/4"
T 2x2, 1/4"
T 2x2, 1/4"
L 5x3, 1/2"(CONT.)
(WELDED TO ANGLE)

IMAGE INDEX + REFERENCES

001 Architect: portrait, on stair structure [11/2020; mg*]
002 First day of excavation [8/2019]
003 Ongoing excavation; guard rail protection [8/2019]
004 Beginning of formwork for footings [9/2019]
005 Pouring concrete for the foundation [9/2019]
006 Stripped foundation wall at rear [10/2019]
007 Continuation of footings, installing re-bar [10/2019]
008 Stepped footing: after excavation it was discovered that the adjacent building's foundation was less deep than expected. The stepped footing prevents underpinning, which is expensive and risky [11/2019]
009 Rear part of foundation completed [11/2019]
010 Welding "skinny plates" with anchor bolts for steel columns to re-bar, to prevent movement when concrete is poured. The proper location is crucial for the geometry of the building's steel frame [11/2019]
011 Fabrication of steel frame at shop [11/2019]
012 Boring for shoring system [12/2019]
013 Installation of soldier beam into boring hole [12/2019]
014 Pouring part of front foundation wall [12/2019]
015 Steel column installation; the shear studs (welded to the column) will be embedded in the concrete block demising wall for lateral stability [12/2019]
016 Installed shoring system w/ rakers, street side [12/2019]
017 First section of corrugated decking installed [12/2019]
018 Partial steel frame erection on rear foundation; the front part requires sidewalk bridging [12/2019; sac*]
019 Partial steel frame erection, rear foundation [12/2019]
020 Pouring concrete footings, street side[12/2019]
021 Foundation wall on street side completed [12/2019]
022 Architect, w/ almost completed foundation [12/2019; jb*]
023 Preparation for west foundation wall pour; before, the excavated soil was used to secure the formwork, here, in the absence of any remaining soil, heavy timber rakers were installed [12/2019]
024 Stripping of formwork, west foundation wall [12/2019]
025 All cleaned up, after completion of foundation [12/2019]
026 Device invented and fabricated by architect, to create conduit in foundation wall for intercom wiring [12/2019]
027 Device, now embedded in concrete foundation [12/2019]
028 Installing floor beams, w/ moment connection [1/2020]
029 Installed steel columns, 1st tier only [1/2020]
030 Corrugated decking on steel ledges; this solution was chosen, so that the depth of the steel beams wouldn't take away too much of the ceiling height [1/2020]
031 Continuing installation of steel framing [1/2020]
032 Delivery of steel; completed foundation [1/2020]
033 Conduit from street to cellar, for main power [1/2020]
034 Timber sheet piling, for sewer/water connection; the trench had to be ~13 ft. deep, due to frost line and permitted "drill in" connection to sewer line [2/2020]
035 Excavation to sewer line [3/2020]
036 Beginning of concrete block demising wall [3/2020]
037 Progressing concrete block demising wall [3/2020]
038 Demising wall: every 4 feet a reinforced bond beam is necessary to prevent cracking of bond [3/2020]
039 Core drilling, through foundation wall, in order to establish water connection [3/2020]
040 Sewer connection trench [3/2020]
041 Completed water connection, cellar [3/2020]
042 Pockets at demising wall, to receive steel beams; blocking by architect, coordinated w/ masons [3/2020]
043 Continuing concrete block demising walls [3/2020]
044 Continuing concrete block demising walls; East wall with structural integration of steel columns [3/2020]
045 Building structure, presentiment of space [3/2020]
046 Street view of building structure; lunch break [3/2020]
047 Building structure, play of light and shadow [3/2020]
048 Steel plate, integrated w/ anchor bolts into demising wall, to receive secondary steel beam [5/2020]
049 Emerging structure, after pandemic work stop [6/2020]
050 Steel structure installation w/ pulley system [7/2020]
051 Delivery of new steel elements [7/2020]
052 Continuing steel installation [7/2020]
053 Lower steel tier, in progress [7/2020]
054 Transition between masonry + steel structure, Street facade/entrance; the last 1-1/2 blocks could not be completed before steel column was set [7/2020]
055 Corrugated decking installation, by architect [8/2020]
056 Steel structure, entrance area; without stair [8/2020]
057 Device built by architect, to pull up secondary steel beams, for installation in stair area [8/2020]
058 Installation device [8/2020]
059 Connection of column to main beam by means of expanding box bolts; the system allows for HSS columns, no stiffener plates were necessary; the connection was crucial for moment frame bracing [8/2020]
060 Installed secondary beams and wall plates for the next upper floor, waiting for continuing installation [8/2020]
061 Installation of upper floor secondary beams [8/2020]
062 Wall plate; main beams stored at cellar below [8/2020]
063 Third floor level; controlled access zone [9/2020]
064 Street view, steel structure [8/2020]
065 Lower tier (up to third floor level) completed [9/2020]
066 First floor ready for radiant heat pipe installation; PVC sleeves are mounted for upcoming plumbing work, so to avoid core-drilling; sleeves made and installed by architect [9/2020]
067 Installed radiant heat system, w/ manifold [9/2020]
068 Ongoing concrete masonry work [9/2020]
069 Pouring first floor concrete slab, preparation [10/2020]
070 Completed radiant heat system, first floor [9/2020]
071 First floor, PVC sleeves, pour stops at edges [9/2020]
072 Detail of floor assembly: radiant heat; sleeves [10/2020]
073 Pressure gauge at manifold; all radiant heat circuits are filled w/ pressurized air: architect to monitor pressure

during concrete pour; if system is damaged, pressure decreases, air bubbles in concrete indicate point of damage, which thankfully never happened [10/2020]

074 Completed radiant heat system; at right side: uncoiler, essential for installation, built by architect [10/2020]

075 Pouring first floor concrete slab, preparation [10/2020]

076 Concrete slab, pouring, in progress [10/2020]

077 Smoothing out concrete slab surface [10/2020]

078 Finished concrete floor, the day after [10/2020]

079 Cutting of steel tread+riser for stringer, out of 8x10" HSS; labor intensive: cut by architect [11/2020]

080 Concrete slab, pouring first floor [10/2020]

081 Template for stair stringer; the geometry needed exact dimensions; executed by architect [11/2020]

082 Cutting and lay-out of main steel stringers; architect in charge for processing at steel shop [11/2020; jc*]

083 Stair stringer template installed on site [11/2020]

084 Cutting steel stringers to size, by architect [11/2020; jc*]

085 Completed stair stringer installation w/ wood treads; now the main circulation system exists [12/2020]

086 Continuation of steel frame, upper tier [12/2020]

087 Continuation of steel frame, upper tier [12/2020]

088 Continuation of steel frame, upper tier [12/2020]

089 Installation of corrugated decking, 4th floor [12/2020]

090 Continuation of steel frame, upper tier, seen from neighbor's rooftop [11/2020]

091 Fixing the stair, attaching mandatory hand rail [12/2020]

092 Street view [11/2020]

093 Stair case trouble: all plumbers said, it can't be done; architect to draw dropped ceiling line at concrete block wall to convince them (and it worked!) [12/2020]

094 Continuation of steel frame, upper tier [12/2020]

095 Delivery of insulation panels for seismic gap [12/2020]

096 Continuing steel frame, upper tier; architect (seen at right) to cut and install corrugated decking and complete stair landing [12/2020; st*]

097 Steel frame, upper tier, street view; sidewalk bridge is installed to complete street-side framing [12/2020]

098 Steel frame, now at the top, south side [12/2020]

099 Corrugated decking at 4th floor, completed [12/2020]

100 Completing steel framing at street side [12/2020]

101 Completed steel framing at garden side; all necessary guard railing in place. The steel beams were designed w/ attachment angles for railing installation [1/2021]

102 Continuing steel frame, street view [12/2020]

103 Ongoing steel framing, to top-out building [12/2020]

104 Secondary steel beam installation, 4th floor [12/2020]

105 Fully decked-out at upper roof, garden side [1/2021]

106 Decking at 4th floor; bulkhead roof missing [1/2021]

107 Top of concrete block demising wall with bond beam; plate pockets for roof beams at bulkhead [1/2021]

108 Radiant heat installation 3rd floor, by architect [1/2021]

109 Radiant heat installation 3rd floor, by architect [1/2021]

110 Manifold installation 3rd floor; in order to heat the floor efficiently, six circuits have to be configured, so that none of the circuits exceeds 200 ft. length [2/2021]

111 Pouring concrete slab, 4th fl. and roof [4/2021]

112 Pouring concrete slabs, top floors; a powerful concrete pump must be used to get material to the top [4/2021]

113 Finished slab, lower roof; black sleeves are for roof's storm water drains [4/2021]

114 Installation of roof insulation, with appropriate pitch towards the available roof drains [4/2021]

115 Pre-task meeting for roof installation [4/2021]

116 Delivery of roof insulation and roof membrane [4/2021]

117 Roof inspection; controlled access zone [1/2021; bh*]

118 Applying first layer of roof membrane [4/2021]

119 Torch-on method, bitumen bonds to substrate; very important, especially for roof edges[4/2021]

120 Steel structure completed; guard rails in place [6/2021]

121 Entrance, air shaft; first light-gauge framing [7/2021]

122 North-West corner; filling in the concrete block [7/2021]

123 Steel substructure installation for stair facade [7/2021]

124 Play of light + shadow at concrete block wall [7/2021]

125 Filling in concrete block at column, by architect [7/2021]

126 Ongoing interior light-gauge framing [7/2021]

127 Ongoing plumbing work [1/2020]

128 Installation of first sheathing, street facade [8/2021]

129 Double-height space at 4th floor: A chain hoist and a platform had to be set up for window installation, south side, by architect [7/2021]

130 Arrival of IGU glass, imported from Germany [8/2021]

131 Installation of first window frame at double-heightspace at 4th floor, on make-shift platform [7/2021]

132 Glass stored on A-frames, built by architect [8/2021]

133 Platform and hoist, ready for window installation; built by architect, double-height space 4th fl. south [7/2021]

134 First glass elements installed, [8/2021]

135 Makeshift-platform at 4th fl. south [8/2021]

136 Progressing glass installation, garden facade [8/2021]

137 Progressing glass installation, garden facade [8/2021]

138 Stair w/ most important plumbing cross-over at ceiling area for waste water lines [9/2021]

139 Makeshift-working platforms, at staircase, built to install plumbing and ceiling framing [9/2021]

140 Upper floor glazing completed at double-height space 4th fl. south; platform/ chain hoist removed [9/2021]

141 Cantilevering structure for installation of upper floor glazing now almost removed, to downwards [9/2021]

142 Cantilevering structure for installation now on 4th fl., ready to install glazing on 4th floor [9/2021]

143 Top floor south facade almost completed [9/2021]

144 Installation of 4th fl. south glazing, in progress [9/2021]

145 Installation of 4th fl. south glazing [10/2021; dl*]

146 Installation of 4th fl. south glazing, in progress [10//2021]

147 Progressing dropped ceiling at stair case [10/2021]

148 Window installation 3rd fl.; rain protection [10/2021]
149 Progressing dropped ceiling at stair case [11/2021]
150 Continuing interior light-gauge framing; window at 3rd floor south installed [12/2021]
151 Start of siding at street facade [11/2021]
152 Rental of scissor lift at street facade to install glazing and siding [12/2021]
153 Street view during facade installation [11/2021]
154 Scissor lift, used by architect [11/2021; st*]
155 Glazing installed on street facade [12/2021]
156 South facade glazing, at 2nd fl. now [12/2021]
157 The scissor lift fits exactly into the given opening, specified by architect beforehand [12/2021]
158 Preparation for installing new window frame for 3rd fl. south; background: interior framing [11/2021]
159 Scissor lift in action, street facade [11/2021]
160 Glazing and waterproofing progressing [11/2021]
161 4h floor glazing installed; chain hoist set up by architect, to lift 450 lbs. water heater [12/2021]
162 Sheathing installation at south facade, stair case; by architect [12/2021]
163 Drywall 4th fl., planning for stair to roof [12/2021]
164 Progressing glazing at street facade [11/2021]
165 Steel framing, exit to roof; by architect [12/2021]
166 First floor, glazing installation, at garden [1/2022]
167 Facade work, street side, close to completion [1/2022]
168 Lower roof, w/ exit to roof annex [3/2022]
169 Main facade at garden, in progress [3/2022]
170 Ongoing interior framing, with integrated mechanical, electrical and plumbing (MEP) work [3/2022]
171 Ongoing framing, w/ integrated MEP work [3/2022]
172 Ongoing framing, here at bathroom, 3rd fl. [3/2022]
173 Ongoing framing, 3rd fl.; stored on floor the curtain wall profiles, mullions and gaskets [2/2022]
174 Ongoing siding installation at street facade [2/2022]
175 Siding rainscreen system during installation [2/2022]
176 Siding rainscreen system, installation on scissor lift, siding strips to fit precisely between windows [2/2022]
177 Siding installation, street view; stair case facade (to the right) is still unfinished [2/2022]
178 Panels for stair case facade, perforation by architect: 3500 holes punched with knockout devices [3/2022]
179 Exit lower roof and window, now all enclosed [4/2022]
180 Plumbing for roof drains: difficulties to integrate the "long sweep" elbows into dropped ceiling (plumbing code requirement); luckily it worked [4/2022]
181 Stair case with drywall [5/2022]
182 Frameless door: header must be installed at a precise position before drywall work; by architect [6/2022]
183 Framing, stair case skylight; MEP integration [6/2022]
184 Start of drywall installation, apartment ceilings [6/2022]
185 Finished apartment framing; drywall on site [6/2022]
186 Finished apartment framing + electrical work [6/2022]
187 Upper stair case landing, w/ drywall/opening [6/2022]
188 Cellar, still without continuous concrete slab [6/2022]
189 Interior framing w/ steel strips for later mounting of medicine cabinet, at bathroom [6/2022]
190 The decision to expose the bottom of main beams at the finished apartment ceilings required fire proofing; they were treated with intumescent paint [5/2022]
191 Siding rainscreen system installation, lower roof [4/2022]
192 Spray insulation, at ceiling, 4th floor [7/2022]
193 Spray insulation, thermal and acoustic barrier [7/2022]
194 Spray insulation at demising wall; electrical installations are finished; intercom/thermostat wiring, too [7/2022]
195 Drywall installation at 3rd floor, in progress [7/2022]
196 Drywall installation at 3rd floor; fully taped and spackled [7/2022]
197 Drywall installation at 2nd fl., ceiling w/ beams [7/2022]
198 Installation of first perforated panel, stair case facade; zinc panels must be isolated from steel substructure, here with waterproofing strips [8/2022]
199 Continuing perforated panel installation [8/2022]
200 Ongoing panel installation, seen from interior [8/2022]
201 Perforated panel installation, from interior [8/2022]
202 Ongoing drywall installation, fully taped and spackled, now being primed; windows protected; 4th fl. [8/2022]
203 Ongoing drywall installation; stair to roof missing [8/2022]
204 Ongoing drywall installation, 3rd fl.; fully taped, spackled and primed [8/2022]
205 Carrying the 450 lbs. water heater to the mechanical room at lower roof level, after delivery guy refused to hoist it with his boom truck to the roof, even though he was paid for it; staircase, 3 floors [10/2022]
206 Hoisting the water heater with a chain hoist to the lower roof level [10/2022]
207 Play of light, through stair facade perforation [9/2022]
208 Installation of remaining perforated panels at stair case facade, on 40 ft. ladder; by architect [9/2022]
209 Completed perforated paneling + siding, street facade, premonition of future street wall appearance [9/2022]
210 Contemporary dialogue w/ neighboring building [9/2022]
211 Sidewalk shed removal after completed facade [9/2022]
212 First floor condition at entrance, street facade [9/2022]
213 Play of light, double-height window, 4th floor [9/2022]
214 Observation that daylight can indeed penetrate the entire length of the stair case at certain days [10/2022]
215 Drywall installation, upper floors; completed [10/2022]
216 Waterproofing treatment for showers, 4th fl. [10/2022]
217 Tile work at bathrooms; glass mosaic tile [10/2022]
218 Tile work at bathrooms; detail at medicine cabinet; tiler's ambition to always establish full tiles [10//2022]
219 Drywall installation, upper floors; completed [10/2022]
220 Drywall installation, 1st floor [9/2022]

221 Drywall installation, 1st fl.: Interior door header; the offset-pivot hinge fits into the mortised imprint [9/2022]
222 Siding, street facade [10/2022]
223 Custom wood window mullions; getting ready [10/2022]
224 Street facade [9/2022]
225 Garden facade; detail with operable windows [11/2022]
226 Garden facade; sheathing at stair case facade completed; waterproofing necessary [11/2022]
227 Garden facade; play od light and shadow [11/2022]
228 Cellar: preparation concrete slab: insulation [12/2022]
229 Cellar slab: preparation: waterproofing [12/2022]
230 Cellar slab: radiant heat installation; insulation, waterproofing, heat pipe installation; by architect [11/2022]
231 Cellar slab, insulation [12/2022]
232 Cellar slab, reinforcement mesh on chairs [12/2022]
233 Cellar slab: radiant heat pipe installation [12/2022]
234 Ongoing drywall installation, 1st floor [12/2022]
235 Drywall at stair case, completed [12/2022]
236 Double-height space, window to garden, completed [9/2022]
237 Upcoming DOB visit in the making [10/2022]
238 East parapet wall, cement board rainscreen [11/2022]
239 East parapet wall, cement board rainscreen with integrated lighting fixtures [11/2022]
240 Waterproofing at lower roof [9/2022]
241 Concrete board installation at North parapet, w/ drain for integrated planter system; by architect [10/2022]
242 Lower roof facade; beautiful light [10/2022]
243 Staining custom window mullions; by architect [11/2022]
244 Installing hardware for custom window mullions for curtain wall system; by architect [12/2022]
245 Installed custom window mullions; by architect [12/2022]
246 East parapet wall, lower roof [12/2022]
247 East parapet wall, lower roof; detail [12/2022]
248 Delivery of steel elements, stair to lower roof [1/2023]
249 Fabrication of steel house number; by architect [1/2022]
250 Installation of house number, at steel beam [12//2022]
251 Finished house number, at entrance header [1/2023]
252 Installation of cantilevering treads, stair to lower roof: originally, it was planned to integrate this stair into 4th fl. apartment; impossible, due to egress code [1/2023]
253 Continuing steel stair installation, w/ additional steel studs in adjacent wall cavity [1/2023]
254 Continuing stair installation to lower roof [1/2023]
255 Continuing stair installation; interconnection of exterior cantilevering edge [1/2023]
256 Continuing stair installation to lower roof [1/2023]
257 Continuing stair installation; interconnection of exterior, cantilevering edge [1/2023]
258 Cantilevering steel stair structure; detail [1/2023]
259 Steel pan for stair, to be filled w/ concrete; detail [1/2023]
260 Cantilevering stair; additional structure conceiled in wall cavity, now covered with drywall, painted [1/2023]
261 Street view from above: Con Edison to install gas line; excavator at sidewalk to dig trench [1/2023]
262 Installation of gas line by Con Edison [1/2023]
263 Delivery of cementious material for stair case [2/2023]
264 Pouring concrete slab at cellar [1/2023]
265 Finished concrete slab at cellar [1/2023]
266 Main stair structure, clad with concrete board, for steel tread installation [3/2023]
267 Continuing tile work; 2nd floor [2/2023]
268 Light-gauge framing, for cellar walls [3/2023]
269 Ambient light, upper stair landing; by architect [3/2023]
270 Surveying conditions for entrance enclosure [3/2023]
271 Surveying conditions for entrance enclosure [3/2023]
272 Drywall work completed, 3rd fl. apartment [2/2023]
273 Stair to lower roof; stair treads filled w/ light-weight concrete; prepared for self-leveling topping [2/2023]
274 Stair to lower roof; stair treads filled w/ light-weight concrete; detail [2/2023]
275 Stair to lower roof; preparation for topping [2/2023]
276 Foyer, 1st floor, still without facade [2/2023]
277 Stair to lower roof: polishing of floor surface [2/2023]
278 Finished self-leveling concrete topping [2/2023]
279 Template: upper landing; stair to lower roof [2/2023]
280 Scissor lift, seen from 2nd fl. apartment; installation of custom mullions, upper floors [2/2023]
281 Installed custom mullions; the peg is conceiling a safety screw; shortened/sanded afterwards [2/2023]
282 Custom mullions at street facade, completed [2/2023]
283 Preparation for self-leveling concrete topping, 2nd floor; broadcast w/ epoxy + sand, by architect [4/2023]
284 Self-leveling concrete topping, application [4//2023]
285 Self-leveling concrete topping, application [4//2023]
286 Diverse items of everyday use move to different places, showcasing the particular ambience [4/2023]
287 Stuccoing of main stair case, riser surfaces [4/2023]
288 Stuccoing of main stair case, west wall surface [4/2023]
289 Stuccoing of main stair case, west wall surface [4/2023]
290 Painting of steel pans, for main stair case, before installation, by architect [4/2023]
291 All steel stair pans installed [4/2023]
292 Installed steel pan for a landing, detail [4/2023]
293 All steel stair pans, filled with concrete [4/2023]
294 Stair case space at night, after closing job site [1/2023]
295 Stair case space, from top landing [4/2023]
296 Installation of 4th fl. powder room vanity [5/2023]
297 Concrete block knee wall for 1st fl. foyer; this wall was originally not planned for; it was required for plumbing features, such as siamese connection, etc. [5/2023]
298 4th floor powder room vanity, in progress [5/2023]
299 4th floor powder room vanity, installed lavatory [5/2023]
300 Play of light, entrance area; entire space up to the roof still in need to be finished [5/2023]
301 Arrival of custom windows for south bathrooms [5/2023]

302 Installed window; 3rd floor bathroom [5/2023]
303 Preparations for top landing of stair case [5/2023]
304 Staircase with direct daylight from south [5/2023]
305 Installation of bathroom window, garden side [6/2023]
306 Ongoing installation of bathroom windows [6/2023]
307 Door saddle template, top landing at 4th floor [6/2023]
308 Play of light, top of staircase landing [6/2023]
309 Installation of continuous handrail, stair case [6/2023]
310 Necessary adjustments to fit door saddle; cutting and chiseling of concrete slab; by architect [6/2023]
311 Installed door saddle; top landing 4th floor [6/2023]
312 Mechanical room at cellar, taking shape; gas meters (wall to the right) not installed yet [6/2023]
313 Lower roof, facade siding w/ outlet wiring [6/2023]
314 Lower roof, facade siding w/ scupper spout for integrated planting trough [6/2023]
315 Lower roof, almost finished; coping missing [6/2023]
316 Daylight striking stair case; detail [6/2023]
317 Street facade; 2nd fl. planter still missing [6/2023]
318 Start of kitchen island installation: blend prefabricated cabinets w/ custom elements; by architect [6/2023]
319 Kitchen island, detail; integration of plumbing knee wall and all kitchen appliances; by architect [6/2023]
320 Polishing of concrete floor finish [6/2023]
321 Touch up of stair case, stucco and paint [6/2023]
322 Handrail detail, 4th floor to lower roof [6/2023]
323 Kitchen island, reveal spacer sheet (dark grey), used as template for stone countertop fabricator [7/2023]
324 Door saddle experiment: cast of epoxy, to continue 1/2" epoxy base between wall and floor [7/2023]
325 Street view, seen from roof, w/ planter framing [8/2023]
326 Glazing, installation behind perforated facade [8/2023]
327 Fastening of glass stops for glazing, detail [8/2023]
328 Glazing, installation behind perforated facade [8/2023]
329 Fastening of glass stops for glazing, detail [8/2023]
330 Foyer w/ access to cellar, used as staging area [8/2023]
331 Entrance area: lower slab for enclosure missing [8/2023]
332 Perforated facade, from interior; detail [8/2023]
333 Cantilevering stair stringer installation, garden [8/2023]
334 Street facade, start of 2nd fl. planter framing [7/2023]
335 Floor grating delivered for 2nd floor balcony [7/2023]
336 Planter framing, seen from 2nd floor [7/2023]
337 Planter framing, 2nd floor; welding in progress [7/2023]
338 Planter framing 2nd fl.; inner frame completed [8/2023]
339 Cantilevering stair stringers, to garden [8/2023]
340 Main stair, w/ finished treads [8/2023]
341 Ongoing window installation at garden side [9/2023]
342 Entrance area: lower slab for eclosure now completed; crucial for enclosure fabrication; installation of window frame for foyer; by architect [10/2023]
343 Ongoing window installation at garden side [10/2023]
344 Preparation for 1st fl. apartment self-leveling concrete; broadcast w/ epoxy + sand, by architect [10/2023]
345 Switching relay, thermostat control for heating [112023]
346 Ongoing kitchen installation, 3rd floor [10/2023]
347 Street facade, w/ galvanized sheet metal planter enclosure [10/2023]
348 Interior door header, fully spackled/sanded [10/2023]
349 Template for bathroom vanity; by architect [10/2023]
350 Installation of kitchen island countertops [10/2023]
351 Kitchen island with stone countertop, detail [10/2023]
352 Ongoing window installation, 1st fl. at foyer [10/2023]
353 Start of finishing light shaft at entrance area [11/2023]
354 Start of entrance enclosure installation [11/2023]
355 Entrance glazing / entrance door installation [11/2023]
356 Ongoing facade installation at 1st fl.; two systems meet at an interior corner, w/ minimal tolerance [11/2023]
357 Cut-out for intercom; integration into entrance glazing; coordination w/ facade and security systems [11/2023]
358 Tiled bathroom w/ vanity, completed [11/2023]
359 Waterproofing for last outstanding facade segment, garden side; installed by architect [12/2023]
360 Facade system for garden side: standing seam panels, to be installed from bottom to top, by architect [12/2023]
361 Street facade, 1st floor, intercom finished [1/2024]
362 Set-up for light shaft finishing at entrance; drywall spackling, sanding and priming [1/2024]
363 Magic daylight moment at stair case [2/2024]
364 Drywalling and finishing of interior at foyer [2/2024]
365 Installation: 30 ft. suspended pendant, entrance [2/2024]
366 Perforated stair facade/light shaft, at night [2/2024]
367 Waterproofing at 1st fl. foyer window wall [1/2024]
368 Finished light shaft, down to 2nd floor; this space had to be finished from top to bottom: only access was via work platforms held by steel beams, which were subsequently covered with drywall [2/2024]
369 Suspended pendant at entrance; end of day [2/2024]
370 Installation siamese connection, etc., knee wall 1st fl.; momentarily compromising waterproofing [2/2024]
371 Waterproofing and foam glass insulation above cellar; by architect [2/2024]
372 Concrete curb and waterproofing at entrance, before installation of stone pavers [2/2024]
373 Finished foam glass insulation/waterproofing [2/2024]
374 Cut stone paving (by architect), staged at foyer [2/2024]
375 Installation of siding at 1st fl., by architect [2/2024; jg*]
376 Installation of stone pavers, at front yard [3/2024]
377 Installation of stone pavers, at entrance [3/2024]
378 Installation of stone pavers, at front yard [3/2024]
379 Drywall installation at mechanical room, cellar [3/2024]
380 Delivery of last self-leveling floor topping [3/2024]
381 Broadcast preparation, for entrance/foyer floor [3/2024]
382 Completion of front yard stone paver installation, with penetrable planting area [3/2024]
383 Set-up for mixing self-leveling concrete topping [3/2024]
384 Pouring self-leveling admixture, at foyer [3/2024]

385 Polishing new floor at foyer/entrance [3/2024]
386 Template for dropped ceiling at entrance area [3/2024]
387 Painting of 2nd floor planter and new railing [3/2024]
388 Planning for closing seismic gap, east demising wall; large gap, based on property line survey [4/2024]
389 Ceiling panel at entrance area, installed [4/2024]
390 Start of kitchen installation, 1st floor [2/2024]
391 Touch-up of planter/railing; by architect [4/2024; jg*]
392 Start of metal paneling at garden, 1st fl.; starting a new system requires time to understand best practice and how to deal with tolerances; by architect [4/2024]
393 Start of metal paneling at garden, detail [4/2024]
394 Stone pavers at front yard, w/ new potting soil [4/2024]
395 Staircase facade, from interior [4/2024]
396 Ongoing metal panel facade installation at garden, only with ladders: scaffolding was too expensive [4/2024]
397 Progressing facade installation, at garden [4/2024]
398 Progressing facade installation, at garden [4/2024]
399 Integration of stair case upper landing window into standing seam facade pattern; by architect [4/2024]
400 Progressing facade installation, at garden [4/2024; jg*]
401 Final painting, at stair case [4/2024]
402 Progressing facade installation, at garden [4/2024]
403 Ongoing facade installation, with corner panel being installed; integration of bathroom windows [5/2024]
404 Installation on 40 ft. ladder; at garden [4/2024]
405 Integration of window strip and siding, detail [5/2024]
406 Play of light, at entrance area [5/2024]
407 Finished garden facade, from neighbor's yard [5/2024]
408 Finished street facade planter, 2nd floor [4/2024]
409 Finished front yard and entrance area [6/2024]
410 Necessary addition of kitchen outlet at 1st floor after electrical inspection; best routing without destroying finishes is often through drywall walls/ceiling [5/2024]
411 Installation of garden stair grating [3/2024; jg*]
412 Painting of the colum at East demising wall, before welding of spacer plate to close gap [5/2024]
413 Finished spacer at East demising wall [6/2024]
414 Trellis system, preparation for installation [6/2024]
415 Rented boom lift, for installation of trellis system and for painting all steel elements w/ epoxy paint [6/2024]
416 Ongoing installation of trellis system [6/2024]
417 Installed trellis system, by architect [6/2024]
418 First vines grow from planter on trellis system [6/2024]
419 Trellis system and planter [6/2024]
420 Street facace at night [6/2024; jg*]
421 Trellis system, street facade, detail [7/2024]
422 Marketing/"open house", 2nd fl. apartment [7/2024]
423 Built-in cabinet for 4th fl. apartment [7/2024]
424 Light from Southern stair case window penetrating through entrance door onto front yard [10/2024]
425 Glassblock ventilation flap at bathroom, detail [6/2024]
426 Play of light through street panel perforation [7/2024]

* All of the above referenced photographs were taken by the author/architect, *except as noted* :

mg Marija Gjorgjievska
sac Syderia Asberry-Chresfield
jb Jeff Brown
jc John Cano
st Saeed Taheri
bh Bethany Herrmann
dl Drew Lowenstein
jg Juan Giraldo

** All photos at the "COMPLETED" section were taken by Albert Vecerka, ESTO, New York

*** The Profile photo (pg. 196) was taken by Evi Abeler, City College of New York

REFERENCES

1 Erhard Volkmann + Eva Maria Volkmann: to my parents, whose valuable input I am appreciating much more today than during the time it originally happened; 1985. I love you!

2 Howard Davis, *"The Culture of Building"*, Oxford University Press, 2006, pg. 108-109: "*In traditional society, the builder combined the functions of design and construction that are now assumed by separate professionals, and the system of apprenticeship taught people to take these functions on in a way that combined thinking and doing. The word architect, when it was used up to the eighteenth century, usually referred not to an architect in the present-day sense of the word but to someone who assumed overall responsibility for design and construction. Almost invariably, [these masterbuilders] came out of the building trades, and their responsibility lay not only in the production of drawings but also in work on the building site ... Their work was in fact an extension of the role of the craftsman."*

3 Mario Campi; in conversation, at ETH Zurich, Switzerland 1991

4 Christian Illies/Nicholas Ray, "Philosophy of Architecture" by Cambridge Architectural Press, 2014

5 U.S. Green Building Council: Green Career Development, seen at: https://www.usgbc.org/professionals/green-careers/commercial-developer, 2019

6 Mario Campi; in conversation, at Campi-Pessina Architetti, Lugano-Paradiso, Switzerland 1995

7 Statistic Atlas, Data from US Census Bureau: Household Types in Manhattan, New York County, 2018, seen at: https://statisticalatlas.com/county-subdivision/New-York/New-York-County/Manhattan/Household-Types

8 Anonymous; in conversation, at Selldorf Architects, New York, 1998

9 *"A201-General Conditions of the Contract for Construction"*, §3.6.1; ACD Guide, AIA, 2017

10 *"The Architect's Handbook of Professional Practice"*, Peter L. Gluck, Wiley Publ. 2013, seen at: https://gluckplus.com/architect-led-design-build/ pg. 555: "*... fear of litigation has had a tendency to drive a wedge between builders and architects, creating an intractable wall between the two players. As architects have limited their representation on the construction site, they have also potentially reduced feedback that fuels good practice, and the ongoing educational and technological knowledge that is critical in the rapidly evolving world of building technology."*

11 Carlo Scarpa *"The Craft of Architecture"*, Hatje Cantz Publ., 2004: *The book describes Scarpa's collaboration with his craftsmen, especially Saverio Anfodillo, and their discussions to construct his designs*

12 Carlo Scarpa, Interview with the New York Times, 1976

13 Ludwig Mies van der Rohe, 1929, seen at: Dietrich Neumann "*The Barcelona Pavilion by Mies van der Rohe: One Hundred Texts since 1929*", Birkhäuser Publishing, 2020

14 TMS 402/602, the Building Code Requirements for Masonry Structures, seen at: https://theme.masonrysociety.org/wp-content/uploads/2022/06/TMS_402-602_Public_Comment_Draft_2021-05-31.pdf

15 Directive 2/75 NYC Department of Buildings, seen at: https://www.nyc.gov/assets/buildings/pdf/di919.pdf

16 Julius Posener, Lecture at Technical University Berlin, "*Das steinerne und das grüne Berlin*" ("*The stony and the green Berlin*"); question and answer session after lecture, 1988

17 New York City Administrative Code: *"Taxable Status of Buildings in Course of Construction,* § 11-209", seen at https://codelibrary.amlegal.com/codes/newyorkcity/latest/NYCadmin/0-0-0-203778

18 *"Richtlinien zum Pflichtpraktikum" (Guidelines for compulsory Internships)*, § 5 of Study and Examination Regulations for the Bachelor's Program in Architecture at the Technical University Berlin, 2019, seen at: www.static.tu.berlin/fileadmin/www/40000062/Dokumente_Formulare/Arch_B_Praktikumsrichtlinie_2019.pdf

19 *"A201-General Conditions of the Contract for Construction"*, §4.2.3; AIA, 1987, seen at: www.nhstateparks.org/getmedia/604e303e-3e3e-4e70-9190-8776dd7334d5/AIA-A201-1987.pdf

20 Adolf Loos, *"Grundsätzliches - von Adolf Loos,"* Vienna, 1930, p. 17

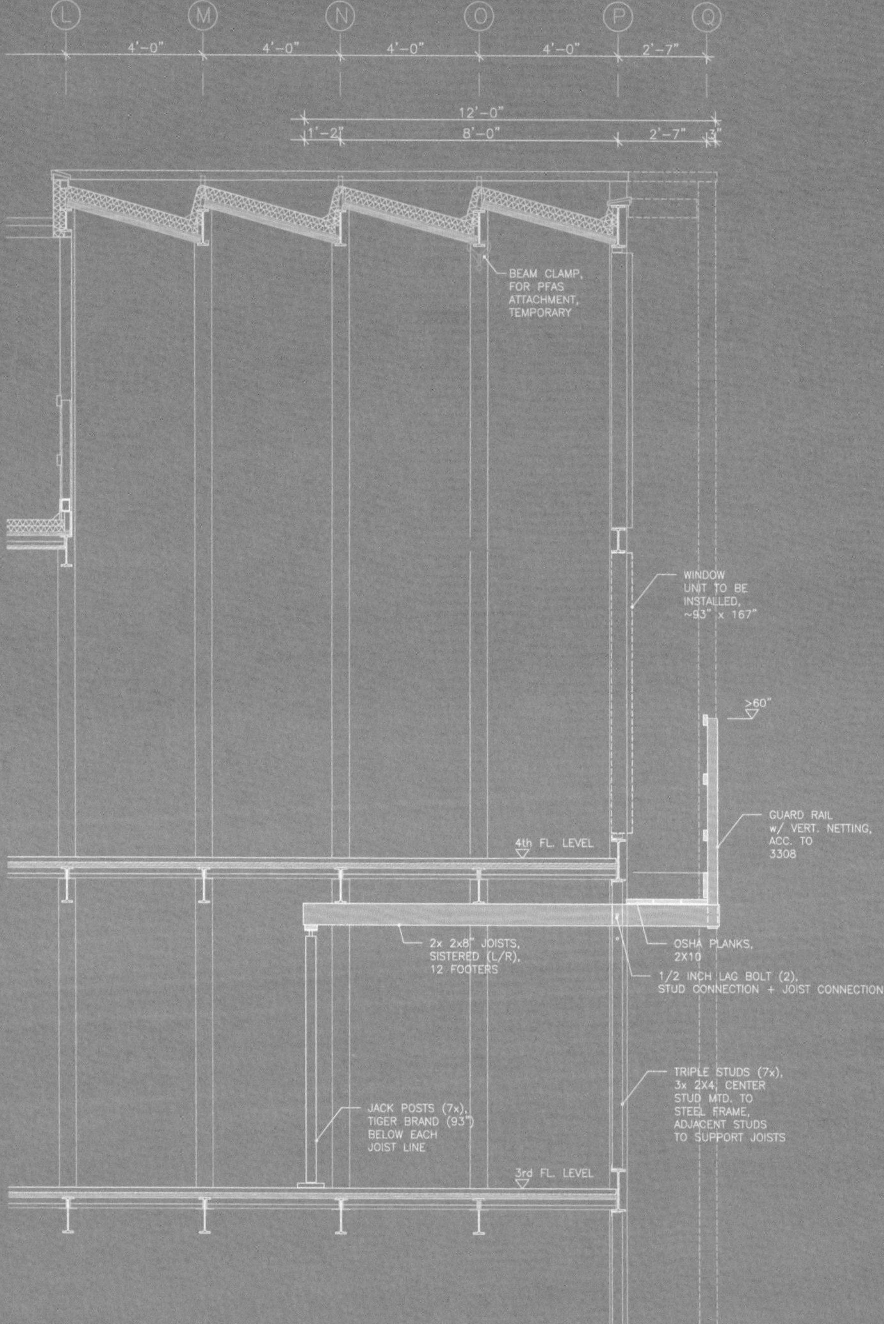

3C / SSP — SECTION: 3RD FLOOR SHORING (SOUTH) DURING WINDOW INSTALLATION

SCALE: 1/2"=1'-0"

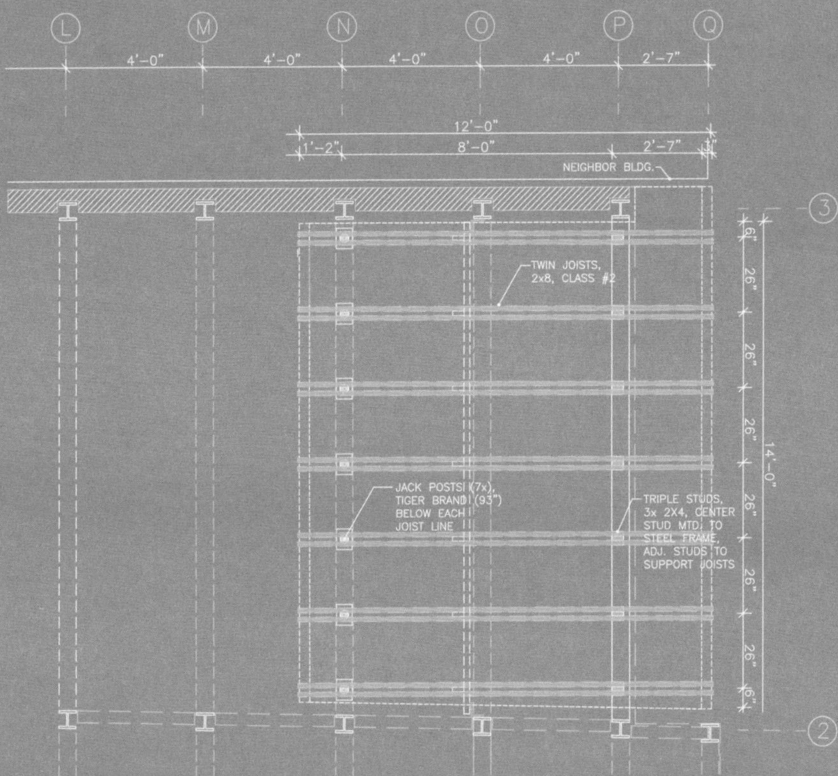

3E
SSP

PLAN: 3RD FLOOR SHORING (SOUTH), BELOW PLATFORM

SCALE: 1/2"=1'–0"

PROFILE

Christian Volkmann is a licensed architect, originally educated in Germany and Switzerland. He is the principal of the New York City based firm of aardvarchitecture. He is a Professor for Architecture at the City College of New York (Spitzer School of Architecture), predominantly in charge of the Technology and the Integrated Design curricula. In order to strengthen both, he has undertaken numerous Design+Build projects with students, among them the Solar Roofpod, documented also in a book by this Publisher.

He has worked in his career, among others, for Josef Paul Kleihues (Berlin), Mario Campi (Lugano) and Annabelle Selldorf (New York).

His body of work focuses on realizing contemporary architectural motives that connect to the whole, resolving the detail architecturally, engaging the motivation of craft with the skills of fabrication and industrial production.

The award-winning project described in this book, the House in Harlem, is inspired by two typical New York City typologies, the Townhouse and the Loft. The building calibrates its contemporary architecture to its landmarked site. Spatial structure, massing and material assemblies reflect on local influences, while interpreting them in a fresh and modern way.

The 4-story building features four apartments, held together by a long stair, which transects the entire depth of the site. Above and below the stair are secondary spaces for the apartments, while the main living space on each floor is uninterrupted, with generous fenestration on street and garden.

The architectural experience in this project is characterized by the sensitive composition and calibration of space, light, form and material.

ACKNOWLEDGEMENTS

It takes a village to raise a house.
It is not possible to do it alone. Many thanks to:

Aaron McIntire, Abba Kloc, Adam Kushner, Akbar Rizvi, Albert Vecerka, Alexandra Bilinski, Amato Pintucci, Anre Morain, Austin Oakes, Avi Golan, Benami Herman, Bethany Herrmann, Bimi Veliu, Brian Silvius, Bridget Smith, Bryan Rojas, Carl Green, Charlton Girard, Cheng-Yi Lin, Christian Cummings, Christoph Pütz, Clement Cheung, Colin Palladino, Corey Rossman, Daniel Kimya, David Arabi, David Johnson, Denis Limanov, Domingo Gonzalez, Eligah Goosby. Emma Waterloo, Erik Nelson, Ermira Kasapi, Erwin O'George, Fernando Franco, Florjan Plepi, Franz Hofer, Gustavo García, Hans-Joachim Arnold, Hugo Alvarado, James Pepe, Jason Aiello, Janusz Malinowski, Joe Gulino, John Cano, John Geraci, John Lakis, John Vlahakis, José Rojas, Josefa Ruiz, Juan Giraldo, Julian Usman, Jun Nam, Justyna Drozdz, Kito Mizo, Kenny Gu, Liz Galan, Luis Carvajal, Luis Espinoza, Luis Ilivasaca, Luis Yunga, Lukasz Wegrzynski, Martin Ossenkopp, Martin Zauner, Matt Pryor, Mateo Penafiel, Matthew Arendz, Michael Aleman, Mike Shamash, Moe Dieng, Muhammad Khedaru, Muhammad Mannan, Nicholas Merhige, Onika Gregory, Paola Tapuy, Pablo Valladarez, Ringo Cheung, Roberto Nasr, Rodrigo Montano, Ron Merhige, Ross Pechenyii, Saeed Taheri, Sebastian Bringmann, Scott Diamond, Sherzod Davronov, Stefan Marshall-Göbel, Stephan Kienle, Steve Preston, Tom Gillespie, Volker Massmann, Wei Zhang, Wensheng Liu, Winnie Chau, Wilton Acosta, Wyatt Kübler

And to my special loved ones: Lynnette Widder, Thilo Widder

① STEEL COLUMN, HSS 4x6
INSERTED AT CMU END,
GALVANIZED, PAINTED

② STEEL BEAMS, W6x16
ALIGNED W/ FRONT OF COLUMN
GALVANIZED, PAINTED

③ CONCRETE BLOCK, 8" WIDTH,
4" HEIGHT, REINFORCED
EXT. EIFS (INSULATION)

④ STEEL PARAPET, HSS 3.5x3.5
12" HEIGHT ON TOP OF BEAM,
GALVANIZED, PAINTED
CAVITY INSULATED w/ XPS INSULATION

⑤ RIGID INSULTION (XPS), w/
DENSEGLASS SHEATHING AND
BLACK WATERPROOFING (GRACE)

⑥ ROOF INSULTION (XPS), w/
TPO ROOF MEMBRANE – 60 MIL
JOHN MANVILLE

⑦ STEEL ATTACHMENT ANGLE, L 5.5x3
WELDED w/ T-PROFILES (8)
GALVANIZED, PAINTED

⑧ STEEL ATTACHMENT TEE, T 2x2
WELDED w/ LOWER ANGLE (7)
MEMBRANE (GRACE) SEPARATED FROM CASSETTE PANEL
GALVANIZED, PAINTED

⑨ TECU CASSETTE PANEL, COPPER OXID
WIDTH: 8" TYP.; GAUGE: T.B.D.,
PERFORATED w/ BACK FLANGES
TO MOUNT IN STEEL TEE (8)

⑩ STEEL FRAME FOR TRELLIS, W 5x19
BOLTED INTO MAIN FRAME (1)
GALVANIZED, PAINTED

⑪ GREENSCREEN TRELLIS, STAINLESS STEEL WIRE TIES,
DECORCABLE "FACADESCAPE",
w/ SUSPENDED LADDER ROD SYSTEM

⑫ LIGHT-GAUGE FRAMED EXTERIOR WALL,
w/ RIGID INSULATION; INTERIOR DRYWALL;
EXTERIOR DENSEGLASS;

⑬ TRESPA RAIN SCREEN, NW14/MT (FRENCH WALNUT)
W/ VISIBLE SCREWS AND FURRING STRIPS ON
MEMBRANE (GRACE, BLACK)

⑭ STOREFRONT GLAZING SYSTEM,
WOOD + ALUMINUM,
w/ CLERESTORY WINDOW

⑮ STAINLESS STEEL HANDRAIL,
WELDED TO PLANTER
w/ WIRE TIES BETW. POSTS,
FEENEY

⑯ STEEL PLANTER, TO MATCH TECU OXID,
REINFORCED AND INSULATED, IRRIGATED
GALVANIZED, PAINTED

⑰ STEEL COLUMN, ROUND 3.5" DIAM.
TO SUPPORT STEEL FRAME ABOVE,
GALVANIZED, PAINTED

⑱ STEEL CHANNEL, CUSTOM, T.B.D.
TRANSITION TO TECU CASSETTE PANELS,
GALVANIZED, PAINTED

3 A503 DETAIL: FRONT, PARAPET
SCALE: 1-1/2"= 1'-0"

6 A503 DETAIL: SIDE, PARAPET
SCALE: 1-1/2"= 1'-0"

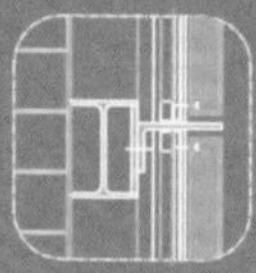

4 A503 DETAIL: FRONT, PANEL ATTACHMENT
SCALE: 1-1/2"= 1'-0"

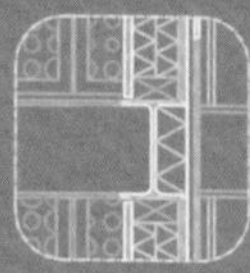

7 A503 DETAIL: SIDE, RAINSCREEN
SCALE: 1-1/2"= 1'-0"

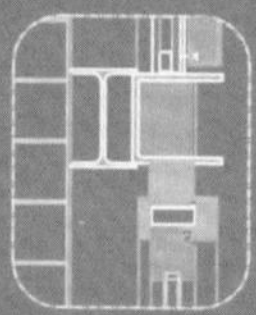

5 A503 DETAIL: FRONT, PANEL BASE
SCALE: 1-1/2"= 1'-0"

8 A503 DETAIL: SIDE, RAINSCREEN
SCALE: 1-1/2"= 1'-0"

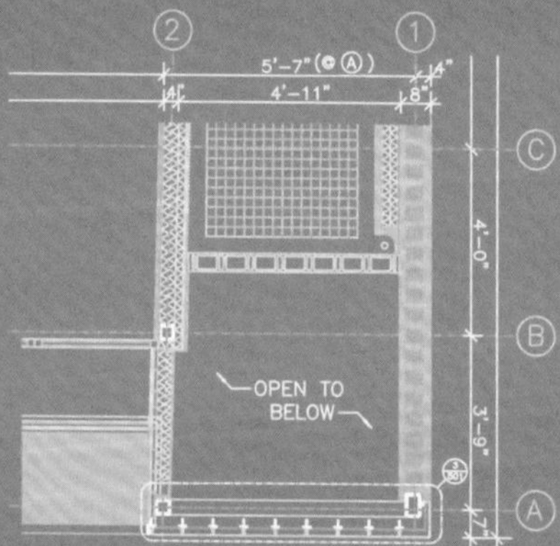

2 A503 PLAN: STAIR, TYPICAL
SCALE: 1/2"= 1'-0"

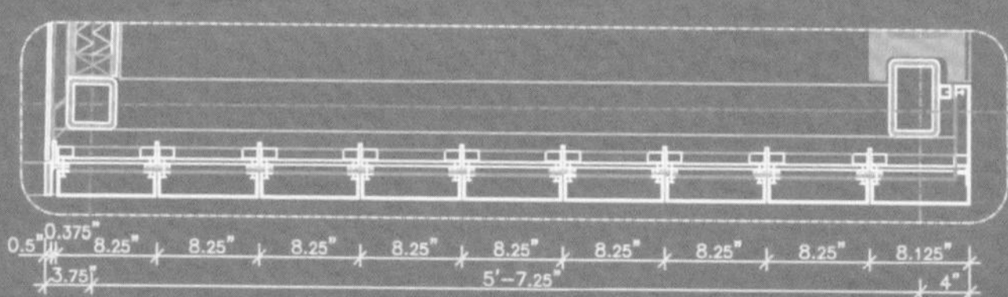

1 A503 DETAIL: PLAN, PANEL ATTACHMENT
SCALE: 1-1/2"= 1'-0"

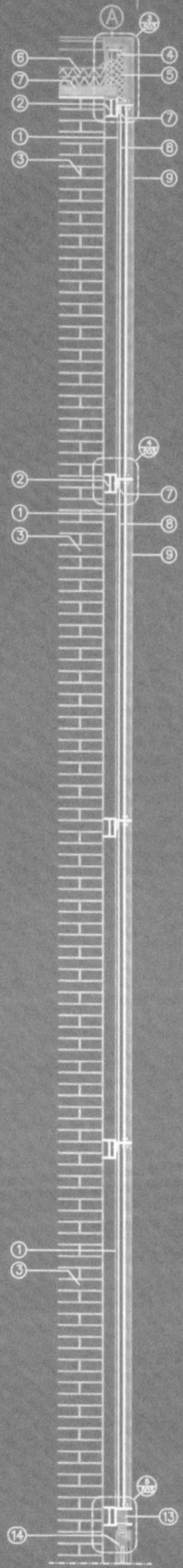

9 A503 SECTION: @ FRONT
SCALE: 1/2"= 1'-0"

11/A503 ELEVATION: NORTH FACADE, STAIR (EXTERIOR)
SCALE: 1/2"= 1'-0"

12/A503 ELEVATION: WEST
SCALE: 1/2"= 1'-0"

13/A503 ELEVATION: NORTH FACADE, STAIR (INTERIOR)
SCALE: 1/2"= 1'-0"

ALTERATION TYPE 1:
NEW CONSTRUCTION/FAMILY RESIDENCE

DATE: SUBJECT:

LOT NO.: 001

DRAWN BY: CV

CHECKED: CV

DATE: 04/05/2015

DEPARTMENT OF BUILDING
FILING SET

TITLE:
WALL SECTION:
STAIR FACADE,
DETAILS

SCALE: 1/2" = 1'-0"

PLAN-NO.:
A-503.00

PAGE: 8 OF 030

BOOK CREDITS

Graphic Design **Juan Pablo Sarrabayrouse**
Editorial Director **Caroline Kipling**
Art Director **Oscar Riera Ojeda**
Copy Editing **Kit Maude**
Editing/Advising **Steve Preston**
Cover Photo **Albert Vecerka, ESTO**

OSCAR RIERA OJEDA
PUBLISHERS

ISBN 978-1-964490-42-7
Published by Oscar Riera Ojeda Publishers Limited
Printed in China

Oscar Riera Ojeda Publishers Limited
Unit 1331, Beverley Commercial Centre, 87-105 Chatham Road South, Tsim Sha Tsui,
Kowloon, Hong Kong

Production Offices
Suit 19, Shenyun Road, Nanshan District, Shenzhen 518055, China

International Customer Service & Editorial Questions: +1-484-502-5400

www.oropublishers.com | www.oscarrieraojeda.com | oscar@oscarrieraojeda.com

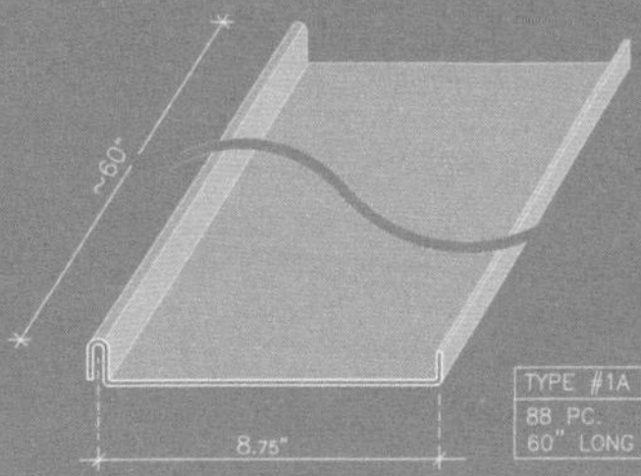

TYPE #1A
88 PC.
60" LONG

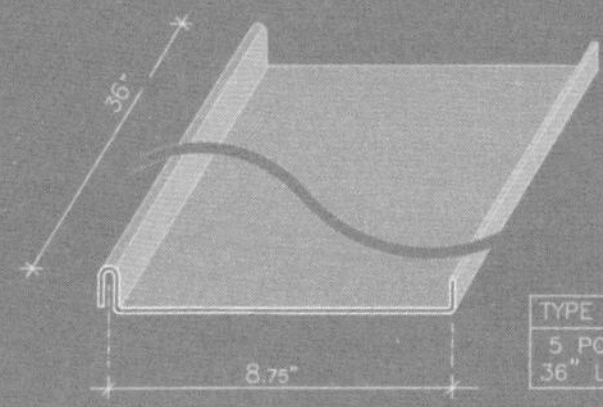

TYPE #1B
5 PC.
36" LONG.

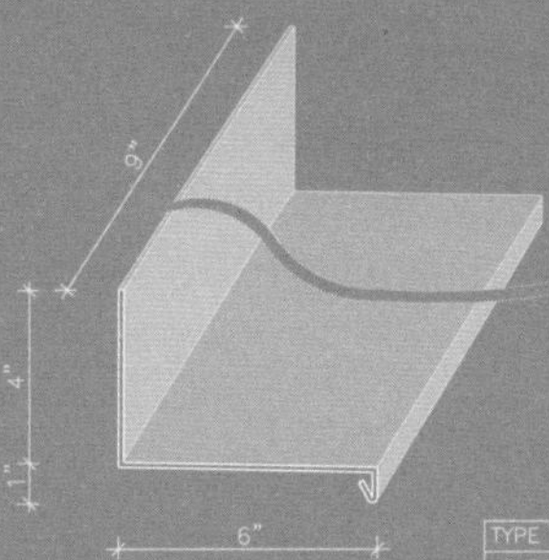

SOUTH WINDOW, UPPER FLASHING

TYPE #3A
1 PC.
9" LONG

EAST WINDOW, UPPER FLASHING

TYPE #3B
1 PC.
12" LONG

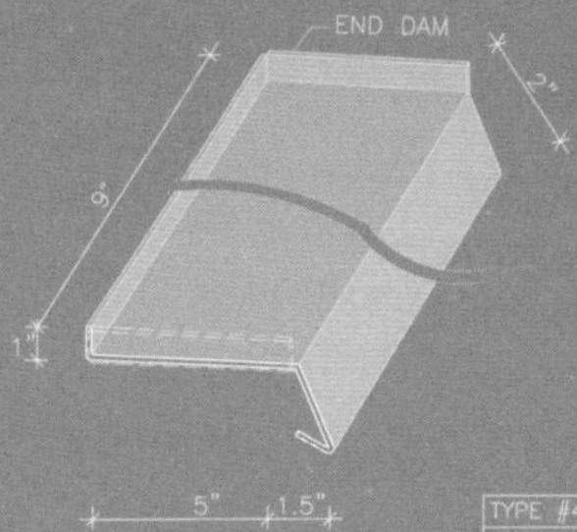

SOUTH WINDOW, LOWER FLASHING

TYPE #4A
1 PC.
9" LONG

EAST WINDOW, LOWER FLASHING

TYPE #4B
1 PC.
12" LONG

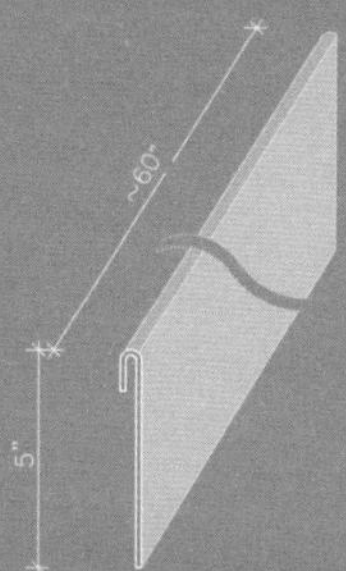

WINDOW JAMB, LEFT

TYPE #5A
11 PC.
60" LONG

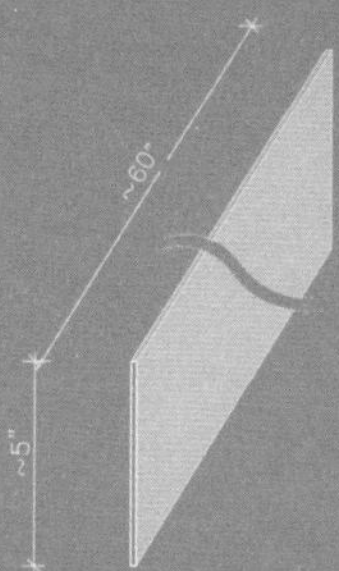

WINDOW JAMB, RIGHT

TYPE #5B
1 PC.
60" LONG

ALTERATION TYPE 1:
NEW CONSTRUCTION/FAMILY RESIDENCE

DATE: SUBJECT:

LOT NO.: 001
DRAWN BY: CV
CHECKED: CV
DATE: 04/05/2015

DEPARTMENT OF BUILDING
FILING SET

TITLE:

WALL SECTION: STAIR FACADE - SOUTH, DETAILS

SCALE: 7/16" = 1'-0"

PLAN-NO.:

A-504.00

PAGE: 8 OF 030

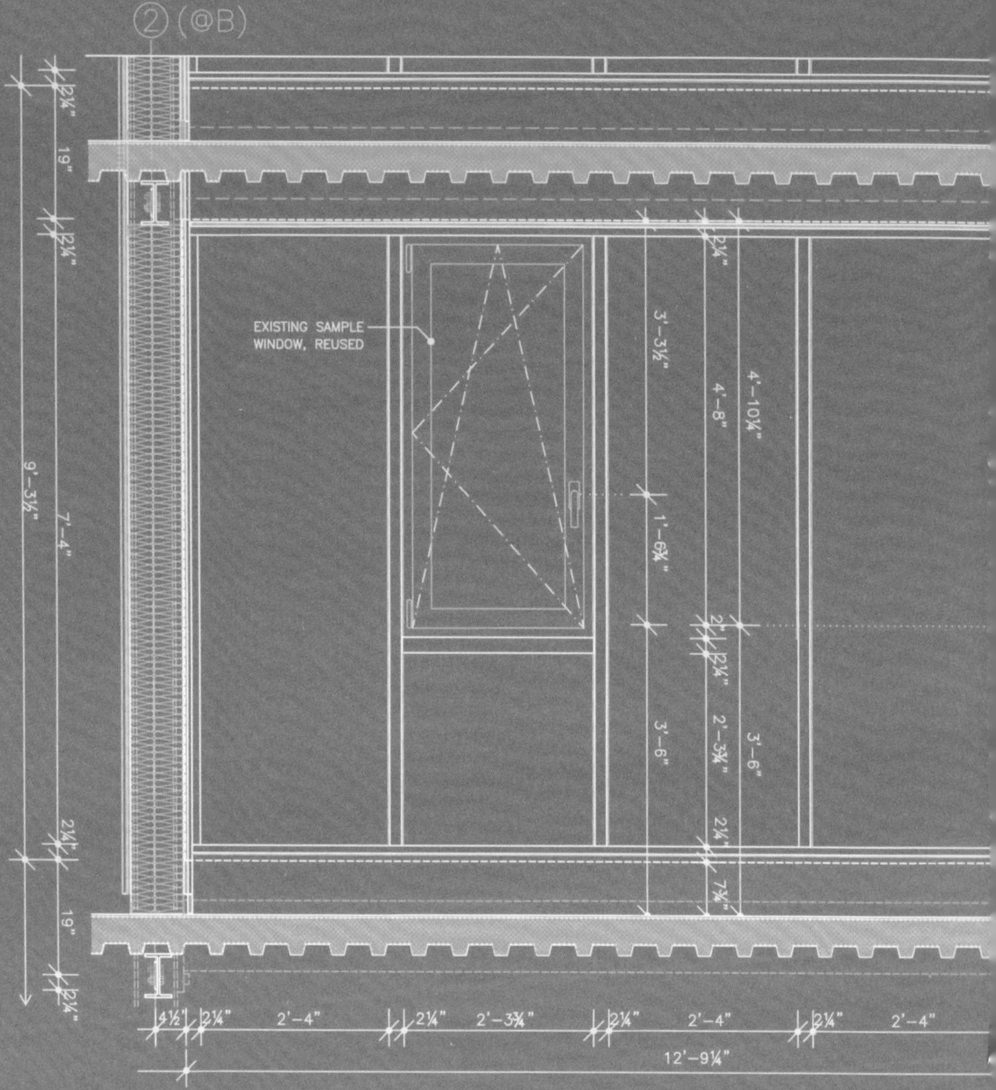

2 W3N ELEVATION, WINDOW WALL SOUTH; INTERIOR
SCALE: 3/8" = 1'-0"

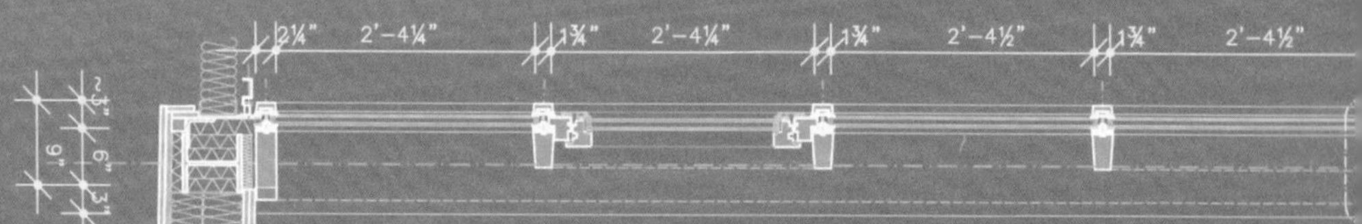

1 W3N PLAN: CURTAIN WALL NORTH
SCALE: 3/8" = 1'-0"